D0432449

Whole Life-cycle Costing
Risk and Risk Responses

Whole Life-cycle Costing
Risk and Risk Responses

Halim A. Boussabaine
Liverpool School of Architecture and Building Engineering
The University of Liverpool

and

Richard J. Kirkham
School of Industrial and Manufacturing Science
Cranfield University

Blackwell
Publishing

Editorial Offices:
Blackwell Publishing Ltd, 9600 Garsington Road, Oxford OX4 2DQ, UK
Tel: +44 (0)1865 776868
Blackwell Publishing Inc., 350 Main Street, Malden, MA 02148-5020, USA
Tel: +1 781 388 8250
Blackwell Publishing Asia Pty Ltd, 550 Swanston Street, Carlton, Victoria 3053, Australia
Tel: +61 (0)3 8359 1011

First published 2004 by Blackwell Publishing Ltd

Library of Congress Cataloging-in-Publication Data
Boussabaine, Halim.
Whole life-cycle costing: risk and risk responses / Halim Boussabaine and Richard Kirkham. — 1st ed.
p. cm.
Includes bibliographical references and index.
ISBN 1-4051-0786-3 (Hardback : alk. paper)
1. Building—Estimates. 2. Building—Cost control. 3. Life-cycle costing.
I. Kirkham, Richard L. II. Title.

TH437.W5 2003
692′.5—dc22 2003015957

ISBN 1-4051-0786-3

A catalogue record for this title is available from the British Library

Set in 10/12.5pt Palatino by Integra Software Services Pvt. Ltd, Pondicherry, India
Printed and bound in the UK using acid-free paper by MPG Books Ltd, Bodmin, Cornwall

For further information visit our website:
www.thatconstructionsite.com

Contents

Foreword by Nigel Dorman, NHS Estates

The UK government has challenged the way its organisations deliver services, and has placed on them a duty to continuously improve in order to provide the services that people require economically, efficiently and effectively. This concept of 'best value' has dominated public sector capital investment policy in the UK since the 1990s. This has been the case particularly in large buildings and civil infrastructure projects such as hospitals, prisons and highways. As a result of the fundamental revisions in public procurement policy that have subsequently taken place, interest in and demand for the use of whole life-cycle costing (WLCC) techniques have risen to unprecedented levels. These policy changes are clearly demonstrated in recent government publications such as *'Construction Procurement Guidance, No 7 Whole Life Costs'* (Office of Government Commerce), which states that 'all procurement must be made solely on the basis of value for money in terms of the optimum combination of whole life costs and quality to meet the user's requirements'. This view is fully endorsed by National Audit Office (NAO) policy and reinforced in their joint guide *'Getting value for money from procurement – How auditors can help'*. Consequently the award of public construction contracts based on simply the lowest capital cost bid is no longer recognised as good practice; best value must be taken into account and thereby WLCC should be fully appraised as part of the decision making process.

Within the UK public sector, WLCC must now be taken into account in all business cases, which aim to justify capital investment in construction. This applies to projects financed by traditional public capital as well as through the Private Finance Initiative (PFI) and Public–Private Partnership (PPP) approaches. The tangible effects of this essential change in procurement can be seen in, for example, the NHS ProCure 21 strategy. ProCure 21 promotes the better use of NHS assets and resources to achieve the right buildings and equipment, in the right place, in the right condition, of the right type, at the right cost (from both capital and whole life points of view), at the right time whilst facilitating effective response to future needs of the service with minimal impact on the environment. The ProCure 21 programme incorporates WLCC models in the tendering process for its frameworks and requires specific models to be completed for each NHS scheme subsequently undertaken by the framework contractors in England. These models have helped the NHS to make significant steps forward in attaining better value for money in capital procurement.

The transition to WLCC-based decision making has been slow and arduous, as this book will demonstrate. The techniques of WLCC have been viewed by many as a complex and highly uncertain science, two descriptions that are perhaps not wholly without merit. In respect of the latter, this book studies in depth the element of 'risk' in WLCC, and presents possible strategies and techniques for dealing with this. However, the continuing research into WLCC will provide us with better models with which to inform the decision making process and deliver best value to NHS stakeholders in the future. This book bears evidence to this, providing examples of the practical applications of the technique and the subsequent benefits that can be obtained.

The authors are to be congratulated on this timely and thought-provoking work, which shows the real value of WLCC, particularly within the economic constraints surrounding public procurement today. I feel sure the book will provide an indispensable reference to practitioners as well as a useful study guide to undergraduate and postgraduate students in the construction and economic disciplines.

Mr Nigel Dorman, BSc, CDipAF, FRICS, FIHEEM
Head of Construction Performance
National Health Service Estates
The United Kingdom Department of Health

Further reading

National Audit Office/OGC. *Getting value for money from procurement – how auditors can help.*
Office of Government Commerce OGC. *Construction Procurement Guidance No 7 Whole Life Costs.*

Useful websites

www.nhs-procure21.gov.uk
www.nao.gov.uk/guidance/topic.htm
www.ogc.gov.uk

Preface

The construction industry has recently experienced a paradigmatic shift in its approach to product delivery and the achievement of customer satisfaction. Where previously the design and construction teams placed a heavy emphasis on delivering buildings at the lowest capital cost, a greater awareness and desire to consider costs over the whole life of the building have prevailed. Clients now want buildings that demonstrate value for money over the long term, and are not interested simply in the design solution which is the least expensive. These changes have led to and highlighted the importance of whole life-cycle costing (WLCC) approaches to the design, construction and operation of buildings.

Rethinking Construction, the government report into the industry, strongly advocated the need to build right first time and every time by considering the long-term costs and economic performance of constructed assets. Additionally, recent health and safety legislation has also placed a specific duty on clients and designers to consider the potential risks of construction, maintenance and operation over the whole life of the building. These drivers, along with the increase in the number of buildings procured under the Private Finance Initiative (PFI) and Public–Private Partnerships (PPP) routes, have led to project stakeholders taking a greater interest in WLCC decision making. So why is WLCC so important?

One of the reasons behind the rise in popularity of WLCC is that it provides a far more accurate assessment of the long-term cost effectiveness of a project than standard economic methods that focus solely on first costs or on operating-related costs in the very short term. WLCC provides vital information on projects such as those procured under PFI, where the consortium requires long-term cost forecasts of service provision that they will be contracted to provide. It also provides the government with knowledge about the anticipated economic liabilities that they will acquire when the asset becomes the property of state. This, however, is just one example of the benefits of WLCC.

Standard cost and value analysis techniques are generally used to quantify and assess the economic implications of a building design. While these techniques do provide a basis for making project cost decisions, they often do not account for many of the parameters, which may affect the actual project value or cost. The existing methods also fail to consider formal decision making processes and risk assessment methods in performing a cost benefit analysis. Investments in buildings are long-lived and as a consequence involve some degree of uncertainty over the life of the building, and the operational and maintenance costs, amongst other factors. If there is substantial uncertainty concerning cost and time information, then a WLCC analysis may have little value for decision making if it fails to account for this. Therefore, it is essential to assess the degree of uncertainty associated with the WLCC results and to take this additional information into account when making decisions.

The book is structured in three parts, each reflecting the importance of WLCC throughout the various stages of the whole life of a building or constructed asset. Although the examples in this book are taken from the construction industry, the intentional aim of this book is to be as generic as possible, demonstrating WLCC with risk assessment as universally applicable to many other capital investment decision making scenarios. The book presents a logical approach to the understanding, development and execution of a WLCC analysis, with the express intention of promoting and inspiring confidence in the process.

Part I deals with fundamentals of WLCC and consists of five chapters, which provide a general background and appreciation of WLCC concepts, whole life risk management techniques and key decision making milestones through the project life. Throughout this book, the terms 'building asset, building facility and project' are used interchangeably and are taken in their widest possible meaning, to incorporate all aspects of the development from inception to eventual decommissioning.

Part II covers aspects relating to WLCC risks and risk responses during the design stage, and consists of five chapters. A key theme in this Part is the concept of integrating service life forecasting, environmental life-cycle assessment and WLCC. Additionally, it also introduces a practical framework for assessing whole life risks and risk responses during the design stage. Part II also includes an innovative framework for developing WLCC budget estimates. The Part concludes with a case study on the practical application of WLCC to the selection of mechanical services. This Part is written in a way that should provide stimulus to the reader to think about WLCC and risk during the design stage, and encourage a holistic approach to design decision making.

Part III considers WLCC issues during the post-design stage of the building life. This includes the analysis of WLCC risks and risk responses during the construction and operational phases. Example risk registers are presented here with guidance on how the analyst should approach and deal with risk. We will also look at some innovative approaches to operational stage WLCC analysis, both for new projects and existing buildings. This Part concludes with a case study example of the application of WLCC in asset occupancy analysis.

Throughout, the book contains a mixture of established theory, practice and innovation relating to WLCC budgeting and risk management. Although we cannot expect to cover all aspects of WLCC, guidance on suitable sources of additional information is provided. Readers who wish to explore some of the issues in the book in greater detail should refer to the list of further reading and references at the end of each chapter.

Halim Boussabaine
Richard Kirkham
Liverpool
May 2003

Acknowledgements

The authors would like to express their sincere appreciation to Mr Ian Hunter and Mr Jeremy Marshall of The Liverpool School of Architecture and Building Engineering, University of Liverpool, who assisted us continually throughout the preparation of this book. Similarly we would like to thank Mr Dennis Bastow and his staff at NHS Estates who have provided us with excellent research support throughout the past 3 years, and particularly to Mr Nigel Dorman for writing the foreword. We would also like to express our gratitude to the following colleagues whom assisted us in delivering this book: Dr Dana Vanier (Institute for Research in Construction, National Research Council, Canada); the members of Whole Life Cost Forum; Dr Stephen J Kirk (Kirk Associates) and Mr Ed Barlett (WS Atkins, Faithful and Gould). The authors would like to acknowledge the many other sources of information, too numerous to list here, which assisted us in delivering this publication.

Finally, we would like to thank our families for their support and encouragement, who no doubt found the ordeal of helping us through the long nights as equally stressful as we did.

The authors affirm that any mistakes and errors in the book are entirely our responsibility.

Ideals are like stars. We never reach them but, like the mariners on the sea, we chart our course by them.

Carl Schurz

Part I
Fundamentals of Whole Life-cycle Costing

1 Towards an Understanding of Whole Life-cycle Costing

1.1 Introduction

Value for money is a concept that is frequently considered when an individual or an organisation is seeking to make a purchase or investment. When acquiring a new car, for example, we may consider the costs of ownership (fuel economy, insurance, maintenance, availability of replacement parts, etc.) when deciding between options. Implicitly then, we consider the long-term costs of ownership in the decision making process. Furthermore, it could be argued that the larger the capital cost of a product, the more important it is to consider these long-term costs. Buildings are a prime example of high cost purchases, yet consideration of long-term costs is not given the attention it deserves. The past 30 years have seen many attempts to encourage a holistic approach to what is in effect 'whole life' cost analysis, but with limited success, particularly in the United Kingdom. One such technique that is currently emerging in the industry is whole life-cycle costing (WLCC).

Whole life-cycle costing is a relatively new concept to the construction industry, albeit based upon the foundations of analytical techniques that have been in existence for some time. It is in essence an evolution of life-cycle costing (LCC) techniques that are now commonly used in many areas of procurement. Like LCC, the primary purpose of WLCC is to aid capital investment decision making by providing forecasts of the long-term costs of construction and ownership of a building or structure. However, unlike LCC it is also a dynamic approach, and can provide up-to-date forecasts on cost and performance throughout the life of the building. Some of the ideas behind the justification for WLCC are synonymous with key issues in today's construction industry:

- *Meeting clients' expectations* Clients (especially in the public sector) now require buildings that are efficient during and after construction. WLCC techniques can demonstrate real cost savings in design solutions
- *Sustainability* Achieving sustainable design solutions relies on the consideration of long-term operational costs and performance of building components
- *Monitoring performance of constructed assets* For example, are PFI/PPP (Private Finance Initiative/Public–Private Partnerships) projects really

cost effective? Only by considering the whole life-cycle costs can this be assessed. Using WLCC also supports benchmarking and key performance indicators

- *Monitoring cost effectiveness of constructed assets* WLCC provides the means by which to constantly review this and base future capital investment on this information
- *Lean construction* By considering long-term cost and physical performance, waste is minimised both during construction and through the life of the building.

The aim of this chapter therefore is to provide a general overview of the fundamental ideas and principles behind WLCC and to demonstrate how it can be of benefit to the construction industry practitioner. The chapter initially examines the history of LCC and its various definitions, moving on to show how WLCC has evolved from LCC as a new and innovative cost analysis tool. The failings of previous LCC methodologies are examined by definition of the innovative aspects of WLCC.

1.2 Whole life-cycle costing: a brief history

Prior to the 1970s, most clients, developers and professionals involved in building procurement made capital investment decisions solely on the basis of capital cost. Outside the construction industry, it was appreciated in some quarters that making decisions solely on capital cost could be folly. They believed that by possibly spending more in capital cost, the long term would realise substantial cost savings when compared with a cheaper alternative. This school of thought was known as 'terotechnology', and it was in effect the beginnings of whole life-cycle cost theory. Within the construction industry, nevertheless, terotechnology was largely ignored. Some of the reasons behind this included an ignorance of the importance of whole life-cycle costs, lack of available data and data collection mechanisms, and the fact that those providing the capital generally had no interest in the subsequent operational costs of the building. In the early 1970s, the term 'cost-in-use' began to

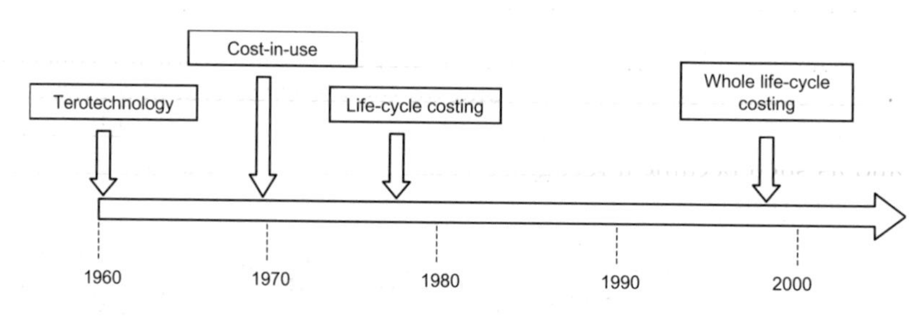

Fig. 1.1 The evolution timeline of whole life-cycle costing.

appear in the industry and the literature. Cost-in-use refers to the expenditure related to the operation of an asset. Although not related specifically to the construction industry, it was recognised that the underlying principles of cost-in-use could apply to buildings and critical structures. What cost-in-use failed to consider, though, was the necessity for accurate future cost forecasting. It became clear then that some kind of technique was required to facilitate this.

It was not until the mid to late 1970s that LCC emerged as a solution to this problem. LCC fostered a wide-ranging approach to cost appraisal, encompassing all perceivable costs from construction through to eventual disposal – 'the whole life'. Using a variety of forecasting techniques, the analyst was able to demonstrate how increased capital cost could be offset by long-term cost savings. LCC sounded good in theory, but the practical implementation within the construction industry did not reflect this. In terms of the enlargement of life-cycle costing theory, the major factor which frustrated its development was lack of good quality cost-in-use and performance data. This proved to be the principal dissatisfaction felt by those who showed some willingness to employ life-cycle costing techniques.

In 1971, the Royal Institution of Chartered Surveyors established the Building Maintenance Cost Information Service (BMCIS) as a method of collecting operational and running cost data. Its main aim was to adopt a single classification system, which could then be disseminated among subscribers in a common demeanour. Although the BMCIS went some way to addressing the implementation problems of life-cycle costing, it did not address the need for a coherent framework and structure in which to deal most effectively with this information.

In 1977, the then UK Department of Industry published *Life-cycle costing in the management of assets* which presented one of the earliest definitions of LCC:

> 'A concept which brings together a number of techniques – engineering, accounting, mathematical and statistical – to take account of all significant net expenditures arising during the ownership of an asset. Life-cycle costing is concerned with quantifying options to ascertain the optimum choice of asset configuration. It enables the total life-cycle cost and the trade-off between cost elements, during the asset life phases to be studied and for their optimum selection use and replacement.'

Since 1977, LCC has become widely reported on, with a diversity of models and techniques existing. In 1983, two eminent researchers in LCC, Roger Flanagan and George Norman, developed a framework for collecting data, which could then be used to build up the life-cycle cost of a project. By 1992, LCC was a familiar concept to building economists throughout the world, and as such became a recognised standard in the UK under British Standard BS 3843 (1992):

> 'The costs associated with acquiring, using, caring for and disposing of physical assets, including feasibility studies, research and development, design, production, maintenance, replacement and disposal; as well as all

> the support, training and operations costs generated by the acquisition, use, maintenance, and replacement of permanent physical assets.'

In 2000, this definition was revised and incorporated into ISO 156868 Part 1 – Service Life Planning which cites LCC as (ISO 2000):

> 'A technique which enables comparative cost assessments to be made over a specified period of time, taking into account all relevant economic factors both in terms of initial capital costs and future operational costs.'

The BS/ISO definition, although authoritative, is a daunting and perhaps vague definition given the plethora of cost items that could be included within each cost category. Principally, the authors believe this to be one of the reasons why LCC is still rarely used to the extent that it was initiated for, although others argue that the lack of quality data is the principal reason. Additionally, the plethora of cost models and definitions associated with LCC has been significant in creating an 'air of confusion' over the subject. Ambiguity and inconsistency were identified in Newton (1991), where consideration of the problem of model classification and the inability to compare models on a like-for-like basis are discussed. Furthermore, the individual perception of the life-cycle model raises many concerns. This is validated in Smith (1999), which highlights how LCC has for some time become an important issue in the overall cost picture, but has not featured in the decision making process to the same extent. This lends weight to the argument in Kirkham *et al.* (1999) that in some respects LCC has remained an academic rather than a practical tool, and that presently the financial burden of implementing an LCC approach outweighs the advocated benefits.

By way of example, consider the application of LCC in other sectors. It has been widely used in the procurement of United States and Australian defence contracts for some time now (Australian National Audit Office 1998; US Department of Defense 1997, 2001). The sheer cost involved in these kinds of projects emphasises the need for LCC; that is, the possibility that significant capital outlay needs to be justified by the longer-term benefits. In some respects, research has shown that LCC has only been applied to projects that have a very high capital cost. In a significant amount of cases, it has been found that ignoring the likely future costs in the conception stage can lead to a significantly more costly endeavour in the future (Smith 1999).

Towards the late 1990s, the concepts of 'whole life costing' (WLC) and 'whole life-cycle costing' (WLCC) emerged. The terms whole life costing and whole life-cycle costing are interchangeable. WLCC is a new term that appears to have been adopted by many building economists involved in the preparation of forecasts for the long-term cost assessments of capital projects. There has been debate amongst academics and practitioners as to whether a difference really does exist between WLCC and LCC. The key emphasis in most of the definitions lies in the implication that LCC is only concerned with the economic life of the building, in other words the period of commercial interest. It could be argued that WLCC forms the attempt by academia and practitioners to overcome some of the problems of LCC. Moreover, it takes into account the costs of running and operating a building over its entire life span – 'the whole life' – as opposed to over a specified period of time, which is a feature of LCC models. Notwithstanding, some have argued that WLCC is simply synonymous with LCC. Others have specified that a difference exists (Bourke & Davis 1999). In this book, the contention is that the concepts are indeed different and to justify this assumption an online survey was conducted (Boussabaine & Kirkham 2000). The authors sought to assess the opinions of academics and practitioners involved in LCC/WLCC, and to establish if the majority of individuals thought that there was real difference between the concepts. This survey forms part of the backbone of the definition of whole life-cycle costing taken in this book.

1.3 Defining whole life-cycle costing

The online survey revealed a broad spectrum of opinion about the difference (if any) between WLCC and LCC. The following are random examples of the responses:

> 'The term...is whole life costing (WLC) and if you believe the old life-cycle costing (LCC) also included all capital and revenue costs for the whole business case/project and not merely elements, e.g. cladding, ceilings etc. then LCC and WLC are...the same thing.'

> 'By rights they should mean the same thing, with the "whole" being superfluous. When one considers a "life-cycle", its wholeness is implied. However, it [may be] possible that when some refer to LCC, they may be referring to the consideration of only costs incurred up to the point when the asset is no longer economically viable and ignoring the issues that relate to asset disposal – which is considered to be part of WLCC.'

> 'If LCC is anything like life-cycle assessment, its completeness is a continuum...In practice, LCAs suffer dramatically from incompleteness, because environmental impacts have to be traced many transactions upstream. This is not a problem for LCC...I can't imagine what the problem would be regarding a small amount of information about the scope of LCC. Australian and New Zealand standards I have seen on LCC do not define WLCC. [Why] would the additional term be required?'

> 'I believe that WLC is simply the modern [equivalent] of LCC. Cost-in-use remains something different. There may be a shift in emphasis to suggest that WLC is not a one-off calculation, but may also be reviewed during the life of the building.'

> '...As I understand it, WLCC goes slightly beyond that to include costs beyond working life – in the case of a building project therefore demolition costs for example would be included...'

> 'In practice, we refer to WLC as the total operating costs of the building, including energy/utilities costs and facilities management elements that relate to the building, such as maintenance and cleaning. LCC refers to replacement building components within the building such as windows, fan coil units, etc. Over and above this are facilities management costs, such as security and catering.'

> '...Theoretically speaking, there is no difference between LCC and WLCC. Each sector adopts a different term. For example, the manufacturing and military [sectors] use LCC, whilst the construction industry may use WLCC and...oil, gas and prime contracting [companies] use through life cost (TLC). However, in the concept of the Private Finance Initiative (PFI), LCC means life-cycle replacement cost which is a part of WLC.'

It would be fair to speculate, after consideration of the points above, whether we really know what WLCC represents. In the absence of any internationally recognised standard on WLCC, it remains a subjective opinion based upon experience, field of work/study and economic standpoint. Of greatest importance is that not one of the respondents in the survey defined WLCC in similar terms; some though pointed out that LCC and WLCC were two of the same thing. In the absence of any national/international standard, who is to say that the above views are at worst incorrect or at least misguided? Although several institutions within the UK are currently working on WLCC-related projects, how can a wider understanding amongst practitioners and academics be initiated, when the research community as a whole is still confused about terms? Naturally it follows that an urgent need exists to define WLCC.

It is the authors' belief that a tangible difference exists between WLCC and LCC. We see WLCC as an evolution of LCC and not a re-invention of the wheel in terms of its association to LCC. Many have rightly noted that a fundamental problem with LCC is the aspect of uncertainty, the risk that is inherent in future forecasting. Some have gone so far as to say that LCC is based on 'guesswork' and 'speculation'. True, we will never be able to state with a very high degree of certainty how a building will perform economically in terms of operation, maintenance and such like. However, it is possible to quantify that risk, to enable stakeholders and decision makers to base capital investment proposals on a basis by which they are aware of the uncertainty in the forecasts. We are also concerned about the fact that the elemental deterioration of building components and the characteristics of the building itself are not integrated into the WLCC decision making process.

In consideration of the results of the survey, and based upon the previous research activities of the various institutions in the UK and elsewhere, the authors advocate the following definition as that of WLCC:

> 'Whole life-cycle costing (WLCC) is a dynamic and ongoing process which enables the stochastic assessment of the performance of constructed facilities from feasibility to disposal. The WLCC assessment process takes into account the characteristics of the constructed facility, reusability,

> sustainability, maintainability and obsolescence as well as the capital, maintenance, operational, finance, residual and disposal costs. The result of this stochastic assessment forms the basis for a series of economic and non-economic performance indicators relating to the various stakeholders' interests and objectives throughout the life-cycle of a project.'

The WLCC assessment components included in the above definition will serve as the guiding principles for this book. WLCC and risk assessment are but part of the overall management of the whole life-cycle of project processes that comprise the art and science of decision analysis. This book addresses the risk assessment-related aspects of decision making in the process of WLCC. The aim is to convey to the reader the fundamental principles of WLCC and risk assessment and the enveloping processes.

1.4 Risk and uncertainty in WLCC

In the previous section, attention has been drawn to the importance of dealing with risk and uncertainty in WLCC analyses. This importance is reflected in the new definition. However, this leads us to a salient point. Is there a difference between uncertainty and risk?

The authors believe there is a great difference. The terms risk and uncertainty are often used interchangeably, although a distinction can be drawn by noting that the concept of risk deals with measurable probabilities while the concept of uncertainty does not. An event contains an element of risk where a probability distribution can be defined. An event is uncertain when no probabilities can be developed concerning its occurrence. Risk refers to probabilities of errors in decisions and WLCC forecasts throughout the life-cycle of a project, or the probabilities of occurrence of events. Risk assessment deals with the likelihood and expectation of possible WLCC outcomes using probability concepts. If computed in terms of the probability of success or failure to achieve the return on investment, the risk is seen as an objective risk. It is an uncertainty when the probability cannot mathematically be indicated but there is enough knowledge to make a subjective judgement about the WLCC decisions. The more explicitly the risk is defined, the greater the possibility for the decision maker to have confidence in using the results of the WLCC analysis.

1.5 Subjectivity in WLCC

The issue of subjectivity and vagueness is also a very important facet of WLCC. Subjectiveness, vagueness and ambiguity (used interchangeably in this book) are different from randomness. Randomness deals with uncertainty (in terms of probability) concerning the occurrence or non-occurrence of an event. Subjectivity, on the other hand, has to do with the imprecision and

inexactness of events and judgements, including probability judgements. Many WLCC decision problems involve variables and relationships that are difficult, if not impossible, to measure precisely. For example, probability judgements about issues like inflation, operation costs, etc. are not always precise in WLCC and often cost analysts use subjective expressions to express their probability judgements. This applies to probability judgements as well as the costs and benefits in many WLCC decision problems. The requirement for high levels of precision may cause WLCC models to lose part of their relevance to the real world by ignoring some of the relevant decision attributes because these variables are incapable of precise measurement or because their inclusion may increase the complexity of the models. Hence, the key to successful WLCC and risk assessment is to build models that require little information – no more than the users can provide. This is a challenge, but it is a challenge that is addressed through the chapters of this book.

1.6 Summary

In this chapter we have looked at the evolution of WLCC and introduced some of the basic principles behind the technique. A working definition of WLCC is proposed, based upon the results of a survey, and we have also briefly introduced the importance of considering risk, uncertainty and subjectivity. In the following chapters we will look in closer detail at the techniques and procedures that are commonly used in WLCC modelling and how these techniques can be modified to cope with risk.

Dealing with risk and uncertainty in WLCC should be the cornerstone of the analysts' approach to WLCC decision making. The uncertainty of forecasting has always been a key problem with practitioners, so by providing information and quantifying the risk element, stakeholders should be more confident in the information that WLCC can provide.

References

Australian National Audit Office (1998) Audit Report – *Life-cycle Costing in the Department of Defence*. Audit Report No. 43, tabled 12/05/1998, Financial Management and Controls Performance Audit, Department of Defence.

Bourke, K. & Davies, H. (1999) Estimating service lives using the factor method for use in whole life costing. *Durability of Building Materials and Components 8: Service Life and Asset Management*, Vol. 3, *Service Life Prediction and Sustainable Materials*. National Research Council Canada Press, Ottawa, Canada, pp. 1518–26.

Boussabaine, H.A. & Kirkham, R.J. (2000) *Survey on WLCC*. Chartered Surveyors Monthly – useful websites for surveyors: www.rics.org/ricscms/bin/show?class=CSM&template=/includes/showcsm.html&id=106

ISO (2000) ISO 15686 Part 1: *Service Life Planning*, International Organisation for Standardisation, Geneva, Switzerland.

Kirkham, R.J., Boussabaine, A.H., Grew, R.J. and Sinclair, P. (1999) Forecasting the running costs of sport and leisure centres. *8th International Conference on the Durability of Building Materials and Components*. National Research Council Press, Canada.

Newton, S. (1991) An agenda for cost modelling research. *Construction Management and Economics*, **9**(2) 97–112.

Smith, D.K. (1999) Total life-cycle cost. *Proceedings of the 8th International Conference on the Durability of Building Materials and Components*. National Research Council Press, Canada.

US Department of Defense (1997) DoD Guide LCC-2, *Life-Cycle Costing Casebook*, DoD Guide LCC-3, *Life-Cycle Costing for Systems*, US Department of Defence, Washington.

US Department Of Defense (2001) Guides/Manuals, DoD Guide LCC-1, *Life-Cycle Costing*, US Department of Defence, Washington.

2 Whole Life-cycle Costing Risk Management

2.1 Introduction

Contemporary developments in the construction industry since the mid 1990s have highlighted the benefits that the whole life-cycle cost approach to investment appraisal can bring to the construction industry. Several reports, including those of Latham (1994) and Egan (1998), have strongly advocated the need to consider the long-term cost of design decisions. Guidance from other agencies has also stressed directly the need for design to encompass a WLCC approach. WLCC has been identified specifically as it gives the analyst the ability to generate tangible evidence of the sustainability of building facilities. It also allows insight into the economic efficiency and performance of the asset.

Recent guidance for projects procured using the Private Finance Initiative (PFI) route advocates the use of WLCC techniques specifically as they provide an assessment of the long-term cost effectiveness of a project. Investments in facilities are long-lived and necessarily involve uncertainty about project life, operating and maintenance costs, and many more factors that affect facilities economics. If there is substantial uncertainty concerning cost and time information, a WLCC analysis may have little value for decision makers. Therefore, it is essential to assess the degree of uncertainty associated with the WLCC results and to take that additional information into account when making decisions. Thus, risk management should form an integral part of the WLCC process. This chapter presents a framework for dealing with WLCC risk management.

2.2 Why has the construction industry failed to embrace WLCC?

Currently, the application of WLCC in the construction industry is still hindered significantly by the lack of standard methods and the excuse of lack of sound data upon which to arrive at accurate decisions. As a result, the output from WLCC models is looked on as unreliable. A government report issued by the Building Research Establishment (Clift & Bourke 1998) on whole life costing identified several factors that presently act as barriers to applying WLCC:

- The lack of universal methods and standard formats for calculating whole life costs
- The difficulty in integration of operating and maintenance strategies at the design phase
- The scale of the data collection exercise, data inconsistency
- The requirement for an independently maintained database on performance and cost of building components.

These barriers might be directly related to the absence of adequate knowledge of WLCC processes and mechanisms. There may also be a lack of willingness from stakeholders to set up appropriate mechanisms to solve these problems. If, for example, all building occupiers were required to submit annual running cost profiles, the risk associated with WLCC techniques could be significantly reduced (Bird 1987). In fact, White (1991) argues the case for 'performance profiles' and in particular, highlights again the requirements for a universal construction data information system. One could argue that a plethora of WLCC models does exist but the common denominator in practical application and development is lack of appropriate information or know-how to use and develop models with existing information.

It seems to be worth noting how both the academic and practical 'schools of thought' in the industry need to get their own houses in order if significant steps are to be taken in the wider applications of WLCC. Newton (1991) in his work in cost modelling procedures highlights the need for a methodological and organised framework for such research activities. The sheer complexity of many models lends little to practical application and in many cases, if not the majority, the lack of available good quality data prohibits further development. In terms of the practitioners, they need to be willing to encourage clients and building occupiers into adopting a more holistic approach to running cost control so that procedures can be put in place to aid all those requiring WLCC cost profiles.

2.3 Why risk assessment in whole life costing?

Combined with WLCC, risk assessment should form a major element in the strategic decision making process during project procurement and also in value analysis, especially in today's highly uncertain business environment. WLCC decisions are complex (the complexity level is usually determined by the scale, funding and financial environment surrounding the scheme, amongst other factors), and usually comprise an array of significant factors affecting the ultimate cost decisions. WLCC decisions generally have multiple objectives and alternatives, long-term impacts, multiple constituencies in the procurement of construction projects, generally involve multiple disciplines and numerous decision makers, and always involve various degrees of risk and uncertainty. Project cost, design and operational decision parameters are often established very early in the life of a given building project. Often, these parameters are chosen based on owner's and project team's personal experiences or on an ad hoc static economic analysis of the anticipated project costs. While these approaches are common, they do not provide a robust framework for dealing with the risks and decisions that are taken in the evaluation process. Nor do they allow for a systematic evaluation of all the parameters that are considered important in the examination of the WLCC aspects of a project. The existing methods also do not adequately quantify the true economic impacts of many quantitative and qualitative parameters.

Capital costs and future costs must be quantified, analysed and presented as part of the strategic decision making process in today's business environment. Cost analysis and value analysis techniques are used to quantify and assess the economic implications of investment in building facilities in general. These techniques have typically concentrated on utilising life-cycle and comparative cost procedures to determine either the lowest initial cost alternative or the highest investment return alternative. While these techniques do provide a basis for making project cost decisions, they most often do not account for many of the parameters which may affect the actual project value or cost (Plenty *et al.* 1999). The existing methods also do not use formal decision making processes and risk assessment methods in performing cost benefit analysis.

It is not surprising that the BRE report (Clift & Bourke 1998) found that clients have a lack of interest and trust in the value of whole cost exercises. This might be due to the fact that WLCC analysis that is not supplemented with risk assessment of all aspects of decision making involved in this process has little value to any decision maker. Risk assessment should be an integral part of the WLCC process. A significant effort is being invested in developing a framework for data collection but data collection on its own, without the correct methods and tools to identify important decisions in the WLCC process and associated risks, is of little value to clients or other interested parties. Clients did not know what to do with information when received and practitioners did not understand the process or the benefits of WLCC (Clift & Bourke 1998).

A framework that uses formal decision making processes and risk assessment of each aspect of the decision to be taken in performing WLCC life-cycle analysis can help owners, design teams and cost planners in making strategic decisions based on analysis results that truly reflect the inherent risks and costs related to the project. Alternative building decisions can be evaluated and compared early in the project development stage. Simply choosing an alternative as being 'better' than the others, based on the traditional approach of life-cycle costing, may not be strictly correct. The confidence of the decision maker's choice depends on the level of uncertainty in the variables and decisions considered. If the WLCC computation results were aided with probabilistic information on the potential costs and risks of the various decisions that are taken throughout the whole life-cycle process of a facility, the confidence in the process of whole life-cycle costing would probably be increased. The subsequent chapters in this book are intended to provide this.

2.4 Data requirements in whole life-cycle costing and risk assessment

Flanagan and Norman (1983) highlighted three fundamental requirements in successfully implementing a life-cycle costing methodology:

- A system by which the technologies can be used: a set of rules and procedures
- Data for the proposed project under consideration: estimates of initial and running costs of elemental life-cycles, discount rates, inflation indices, periods of occupancy, energy consumption, cleaning and the like. The data required to carry out WLCC analysis can be derived from a range of possible sources (Bennett & Ferry 1987):
    - Direct estimation from known costs and components
    - Historical data from typical applications
    - Models based on expected performance, average, etc.
    - Best guesses of the future trends in technology, market application
    - Professional skill and judgement.

All these factors have some bearing on the quality of data that is collected and how it is used in modelling and decision making processes. Whilst WLCC is now becoming widely used as a valuable tool in the design process, probably two key factors have undersized its potential impact (Flanagan *et al.* 1987; Bird 1987):

- A suspicion that life-cycle cost estimates are in some sense inaccurate or based merely on guesswork
- The absence of sufficient and appropriate cost and performance data.

2.4.1 Data sources

It has been highlighted how important the data and its composition are to WLCC, but where can this data be obtained? Ferry and Brandon (1991) highlighted six main outputs:

- Technical press
- Builder's price books
- Information services such as the Building Cost Information Service (BCIS)
- Government research literature such as from the National Economic Development Office (NEDO)
- University research
- Technical information services.

Flanagan and Norman (1983) defined these into four subgroups:

- Manufacturers' data
- Suppliers and contractors
- Modelling techniques
- Historical data.

The availability of data has a significant impact upon the types of modelling techniques available to the analyst. For instance, monthly data for energy costs can be used to develop time-series forecasts, annual data can be used to model probabilistic distributions, etc.

2.4.2 *Manufacturers' data*

A unique aspect of the building process is its individuality and specialist requirement. Many building materials and indeed processes are subcontracted out to individual specialists. These specialists as a rule will have detailed breakdowns of the life-cycle of the product, its material components and its performance characteristics.

This data can also be obtained from other authorities that are responsible for testing the integrity and materials for construction. The British Board of Agrément is a UK government testing body which carries out independent testing of materials used in the industry. Materials that meet a set specification and performance are issued with Agrément certificates, which give details on service lives and other critical information. The Building Research Establishment also carries out testing on materials and can be a useful source of information. Furthermore, the American Society of Civil Engineers regularly publishes papers on building material life spans for use in cost profiling (Ehlen 1997).

2.4.3 *Forecasts from models*

In the absence of any historical or suppliers' data/feedback, models can be used as a way to analyse the WLCC implications of particular design decisions or choices of materials. The concept behind modelling is to facilitate and introduce a higher degree of accuracy in the estimates made by cost analysts when drawing up life-cycle cost profiles. Some dispute the validity of model forecasts but one school of thought advocates that simulated forecasts are as good if not better than historical data due to the following reasons (Ferry & Brandon 1991).

- Historical data by definition relates to the past, whereas simulated data refers to the future. The argument is that for maintenance and servicing costs, data recorded previously would be a poor guide as in the future more sophisticated facilities management techniques and higher quality products and reliability would provide for a different cost structure.
- It is very fortunate to obtain historical data that has not been trained or recorded for other purposes. Typically, in dealing with life-cycle costs, data may be required about a particular element but data may only be available generally for a whole group of elements. Simulation could provide this data in the required format.
- It is also believed that historical data can be inaccurate if those job-sheets and other forms filled in on a 'Friday afternoon' are maintained in a slipshod manner.

These idiosyncrasies can be dealt with, however, using a variety of statistical techniques to reduce data noise, data outliers and other characteristics which are likely to incorrectly skew the results. For example, outliers might be determined using box plots; missing or erroneous data sets are corrected using local means (averages). However, a thorough knowledge of the data used in the modelling should be acquired to ensure that the techniques used are valid and not likely to introduce biases into the modelling process.

2.4.4 Historical data

Historical data can be obtained from a variety of sources such as the BMCIS, clients and building occupiers and in some cases the design team themselves. The value of historical data is relevant in that the values of initial capital cost and subsequent running cost can be categorised for certain groups of element in the building and this comparison can then be used to identify the elements which will benefit from a life-cycle cost approach. For instance, if a building element has a high initial capital cost and then subsequently low maintenance and running costs, a life-cycle approach would gain little use. However, elements with significant running costs could, through design change, for instance, benefit from life-cycle cost savings.

2.5 Specifying a comprehensive set of objectives and measures for each WLCC component

The authors advocate, for this purpose, the use of operational research (OR) and methods described in risk assessment and decision making in business and industry (Pilcher 1992; Koller 1999). OR methods consist of a number of well-defined scientific steps:

(1) Formulation of the problem, establishing the objectives and any constraints that may apply
(2) Building a model that represents the system under analysis
(3) Using the model in order to obtain a solution to the problem
(4) Comparing a solution obtained by means of the model with that in current use
(5) Evaluating the results and monitoring the performance of the system through changing conditions.

The above steps must be applied to every aspect of the WLCC, as defined in Fig. 2.1. Therefore, the interactive steps that are involved in this process are:

2.5.1 Service life

The prediction of component service life is a very important aspect in WLCC assessment. One such methodology currently in use is the factor method. The ISO/CD 15686-1 factor method for the estimation of the service life of

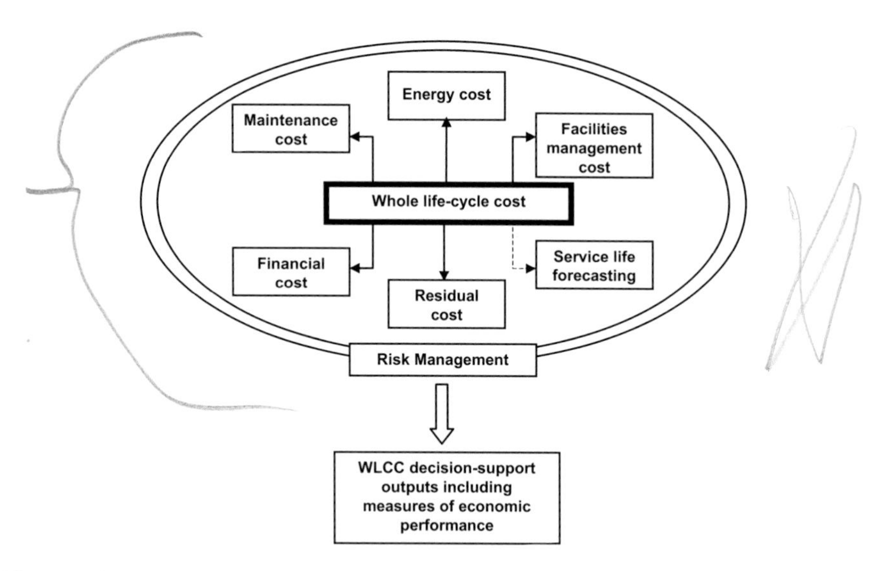

Fig. 2.1 Components of a whole life-cycle cost analysis.

components or assembly (facility) under specific conditions treats the service life as a deterministic value. In reality the service life has a big scatter and should be treated as a stochastic quantity. Here the authors advocate that at early stages of the project (design) this approach should be used but in conjunction with uncertainty and risk assessment for each of the involved factors. The objectives that need to be assessed include:

- Assumptions about the service life of the major facility components
- Risk of failure of components
- Quality of fabrication and production
- Assumption about the updating and maintenance management plans
- Cost constraints
- Assumptions about frequency and time intervals of maintenance and replacement.

At operation stage the requirements are different and necessitate a completely different method of assessment. The process here is concerned with the prediction of the remaining service life of the facility components and the forecasting of the rate of their deterioration and the maintenance budget required to bring the state of the facility to an operational standard. For this purpose the authors advocate the use of the Markov theory (explained in detail in Chapter 5). The objectives that need to be identified and assessed at this stage include:

- Survey condition of the existing facility components
- Assumption about the remaining service lives of components
- Updating budget requirement
- Priority of components updating – critical components
- Quality of maintenance and replacement components

- Assumption about time-lag replacement or maintenance delays
- The effect of delayed maintenance on budget and deterioration of facility
- Evaluating the economic viability with a view to disposal.

2.5.2 *Obsolescence and end of use*

Societal and business environments are characterised by rapid changes in technology and user patterns, which also result in changes in the requirements for infrastructure. Changes in use, trends in fashion (this is more likely to be in housing), and the emergence of new technologies (mainly mechanical services and IT) will have a direct effect on the life expectancy of components and facilities in general. Facilities should be designed and constructed with flexibility in mind to allow future adjustments for anticipated and/or likely changes. The objectives that should be identified and assessed include:

- Intangible costs, such as access disruption to building activities or other costs borne by other than clients, which will also be considered under this section
- Birth, growth and death of firms (business). This can be achieved by projecting the number of years that a business is viable. This will have a direct effect on the functional life expectancy of a facility. This assumption can be used for preparing a life-cycle cost analysis
- Physical obsolescence
- Economic obsolescence
- Technological obsolescence
- Functional obsolescence
- Legal and social obsolescence
- Disposal and decommissioning methods
- Demolition, retain or refurbish options
- Asset reuse or recycle of components
- Waste disposal implications
- Site and land clean-up procedures.

2.5.3 *Capital costs*

Returns on invested capital costs are essential in making decisions on investment scenarios. This requires a combination of knowledge about the investment in question, skill for analysis and elicitation of decisions from the existing information, experience and judgement. The capital cost for acquiring a facility will not be known with certainty until the facility is developed and handed over for operation. Hence, the information required for carrying out whole life-cycle cost and economic viability analysis relies on the availability of previously documented cases and speculative assumptions. The capital cost objectives that need to be assessed include:

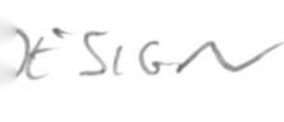

- Land acquisition cost. The location, and land viability may have a direct effect on the whole life cost and life expectancy of a facility

- Predesign costs. The amounts of time and quality of information generated (development of the brief and facility specification) at this stage have great consequences on the quality and operation of a facility. The investors have a good opportunity to optimise the whole life cost of a facility through the selection of component and functional flexibility. Ideally, the issues relating to obsolescence should be investigated, accounted for as costs at this stage
- Design costs. The quality of design in terms of error, detailing and buildability will have a direct effect on the cost of production and operation. A high quality building might also require higher costs in use in order to maintain its high aesthetic quality in use (Ashworth & Hogg 2000)
- Development and production costs. The quality of workmanship is directly related to the level of maintenance. It is important to ensure that quality control is in place to ensure sound construction practices are used
- Fees
- Risk costs
- Financial costs, tax, interest, etc.

2.5.4 *Operational costs*

Operational costs are less certain as the time span increases due to uncertainties in energy costs, maintenance, fees, staff and regulatory changes. It is important to view operational cost estimates in their holistic state; several qualitative factors will have an important effect on the total operational costs. The operational cost objectives that need to be assessed include:

- Factors which contribute significantly to the total operational costs
- Optimum balance between capital and operational costs
- Operational risk management systems
- Optimum asset cleaning procedures
- Optimum waste management procedures
- Optimum utilities management procedures
- Optimum staffing level
- Minimum disruption due to denial use of the asset.

2.5.5 *Maintenance costs*

The costs and priority of required maintenance, rehabilitation and replacement can be obtained from historical data but base cost estimates have to be supplemented with expert opinions in order to perform whole life-cycle analysis and risk assessment. The maintenance cost objectives that need to be assessed include (some are from Kirk & Dell'Isola 1995):

- Performance indicators for the assessment of maintenance costs
- Remaining service life of facility components
- Frequency and replacement costs
- In-house or subcontracted maintenance

- Selection of exterior and interior materials and surfaces
- Selection of light fixtures with minimum routine repair and replacement requirements
- Type of preventive maintenance programme.

2.5.6 Financing costs and revenues

The objective here is to deal with WLCC input parameters of discount, inflation rates, taxes, expenses, etc. Critical analysis of investments must include both initial and ongoing costs and returns over the period of the investment. This will allow stakeholders to compare different options and decide which offers the best return for the investment. Usually discount rate is used for computing the value of future revenues. This includes a large degree of risk return. For example, if the discount rate is set too high or too low then future costs may appear insignificant; this could result in high operational costs and capital costs, which will discourage investment. Also, if inflation is different from the selected rates this may lead to inappropriate investment choices. The financing cost objectives that need to be assessed include assumptions about:

- Inflation rates, interest and taxes
- Level of returns and risks
- Optimum discount rate
- Economic activity. This has a direct effect on the economic obsolescence of facilities
- Level of risk financing
- Cash inflows versus outflows
- Different rates, time periods and cash flows.

2.5.7 Asset characteristics

The characteristics (i.e. physical and functional) of new or existing facilities are very important aspects of WLCC computation. The research community has largely ignored this aspect of WLCC. For example, a relationship may exist between building function and mechanical services costs, a particularly important feature of modern facilities. Little research has been published with regard to the impact of building characteristics on WLCC. Experience shows that an indirect link exists through many aspects, including energy costs for example. A poorly insulated building will consume more energy, thus increasing WLCC and possible downtime costs in maintenance (Department of Industry 1977). The characteristics that should be assessed and included in the computation of WLCC include:

- Layout and location
- Functionality
- Construction technology
- Gross floor area
- Number of storeys and storey height

- Glazing area
- Occupancy (m^2/person)
- Shape of the facility
- Aesthetics
- Energy-saving measures
- Quality of components
- Type and quality of public health systems
- Type and quality of superstructure building fabric
- Type and quality of internal fabric
- Type and quality of electrical and mechanical services
- Extent of site works.

2.5.8 Economic performance measures

The procurement of building facilities involves a variety of decision makers who decide on alternatives that generate capital and ongoing costs during a project's life. These capital costs generate value for different stakeholders and potential for returns to the project owner which should be durable over the life-cycle of the asset. Conventional investment appraisal techniques, which focus on cash flows represented by the costs and expected returns of a project discounted to a common base period, do not reflect the total value of capital expenditure choices which include intangible and no-monetary benefits as well as reduction of future costs and financial returns (Plenty *et al.* 1999). Usually these economic measures are not supported by any risk assessment analysis. Therefore, economic performance measurement in WLCC is very important for decision makers to evaluate and allocate identifiable value from capital and continuing costs to relevant stakeholders in the life-cycle of a facility. This will allow the consideration of different stakeholders' objectives in the assessment of the WLCC. The objectives that should be assessed under this heading should include:

- What type of performance indicators should be used to aid in the selection of alternatives
- The boundaries of these indicators, i.e. minimum and maximum values that the stakeholders are prepared to work to
- The best measures of performance in terms of WLCC outputs
- Mechanisms for WLCC benchmarking
- Measures for mitigating economic risks.

2.6 A framework for whole life costing risk management

WLCC risk management is one of the important issues facing building assets executives today. As spending on building assets rises, asset owners become increasingly worried about WLCC optimisation throughout the life span of facilities; consequently, they become highly vulnerable to the risk of operational costs. Usually, when decision makers are faced with an investment choice under uncertain conditions, their main concern is to avoid projects whose actual economic outcome might be less favourable than what is acceptable, resulting in the risk of missing out on potential investment opportunities. Thus, the objective of WLCC risk management should be to assist decision makers in evaluating whole life alternatives so that investment success is maximised. Usually traditional methods are used to optimise this process. However, traditional approaches to risk management have failed miserably because of their demand for mysterious statistical data that the end user does not have (Koller 1999). The key to successful WLCC risk-process and risk modelling is to build a WLCC framework that requires from the user nothing more than they presently can provide. This can be a challenge that can be addressed through the use of a variety of techniques. That is why it is important to use a combination of risk management techniques (depending on the stage of assessment) for risk assessment in WLCC, ranging from simple deterministic approaches to uncertainty assessment (e.g. sensitivity and break even analysis methods which are easy to use and understand and require no additional methods of computation beyond the ones used in LCC analysis), to very sophisticated methods based on probabilities, artificial intelligence (AI) and a hybrid of both techniques.

There has been a great emphasis on the techniques that are used to model risks but there has been little work on the integration of the whole process of risk identification, quantification, response and management strategies. We believe strongly that these interlinked processes are essential ingredients for any successful risk framework in the life-cycle of assets. Figure 2.2 shows the interaction between these processes.

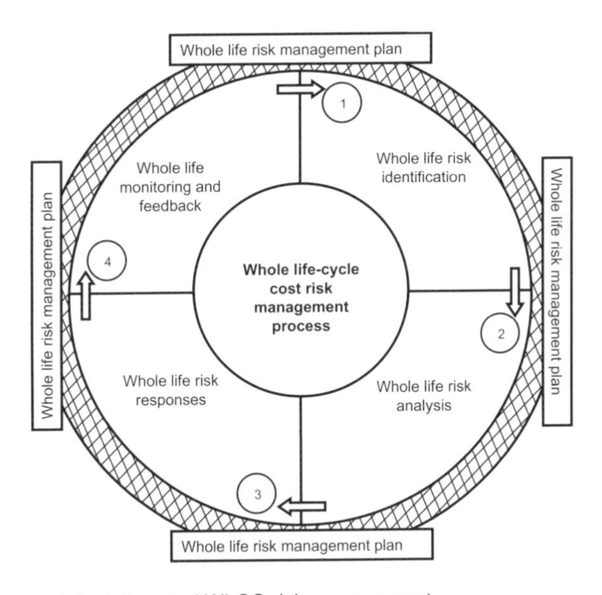

Fig. 2.2 A framework for integrated WLCC risk management.

The whole life-cycle risk process for each stage of the project might be divided into five interactive steps, as shown in Fig. 2.2. This framework does not emphasise any one particular component (though identification is deemed to be the most important) of the WLCC risk management process but concentrates on the sequential and iterative linkage of the five components to make up the entire process of WLCC risk management. If WLCC risk is addressed in this fashion it should enable stakeholders to smoothly move forward and backward from one component to another by identifying and understanding the possible courses of action in the different steps. Such a framework should provide whole life risk managers with a comprehensive view of the overall WLCC risk management strategies. The framework, integrating the five iterative steps, is presented in Fig. 2.2 and explained in the following sections.

2.6.1 Whole life risk identification

The process starts with a qualitative stage that focuses on identification of risks related to each of the whole life-cycle processes. Risks that are unidentified and not quantified are unmanaged risks that can have a significant negative outcome on projects and organisations. If any of the unidentified risks occur at any stage of the project life-cycle, this may have serious consequences on stakeholders' financial status. Hence, perhaps the most important step in the whole life-cycle risk process is the process of risk identification. The quality of this process has a direct effect on the quality and accuracy of risk analysis, quantification, and development of risk strategy responses, and on the management of risk throughout the life span of projects. The output of risk identification will inform the second quantitative analysis process that focuses on evaluation and assessment of risks associated with each aspect of the life-cycle span of projects.

2.6.2 Whole life risk analysis

Several methodologies are available to deal with WLCC risk analysis. The techniques that can be used in WLCC risk assessment decision making might be summarised as deterministic, probabilistic and AI. Deterministic methods measure the impact on project outcomes of changing one uncertain key value or a combination of values at a time. In contrast, probabilistic methods are based on the assumption that no single figure can adequately represent the full range of possible outcomes of a risky investment (Fuller & Petersen 1996). Rather, a large number of alternative outcomes must be considered and each possibility must be accompanied by an associated probability from a probability distribution, followed by a statistical analysis to measure the degree of risk. Using a deterministic approach, the analyst determines the degree of risk on a subjective basis. AI methods differ from the above approaches and use historical data to model cost and uncertainty in WLCC analysis. None of these techniques can be applied to every situation. The best method depends on the relative size of the project, availability of data and resources, computational aids and skills, and user understanding of the technique being applied. We have not provided here a detailed explanation of all the risk analysis methodologies as these are adequately discussed in Chapter 5.

2.6.3 *Whole life risk responses*

Developing responses to reduce WLCC risks is the third step in the integrated WLCC risk management framework. Once the building assets and the many different risks and threats to which they are exposed are identified and quantified and the related life-cycle vulnerabilities assessed, necessary steps should be taken to ensure that the entire investment is protected from all sources of external and internal threats. Thus, the third stage is concerned with the identification of strategies that mitigate the effect of anticipated threats to the greatest extent possible. This should be based on the following universal rules: risk avoidance, risk reduction, risk absorption and risk transfer. The various risk responses that may be implemented to mitigate WLCC risks are explained in Chapters 9, 11 and 12.

2.6.4 *Whole life risk management plan*

Following the identification, quantification and development of risk responses, the related vulnerabilities of building assets need to be determined and planned for. This provides the basis on which risk management plans and decisions are made. The risk management planning process is concerned with putting in place the procedure for:

- What response actions are needed
- When these response actions are needed
- How these actions are implemented
- Who is responsible for the implementation, control and monitoring of the actual progress of risk responses and management strategies that have been developed to deal with the identified risk.

2.6.5 *Whole life risk monitoring and feedback*

The issue of risk monitoring is essential for ensuring effective implementation of risk control measures. Active risk monitoring ensures that effective response measures to manage the risks are appropriately implemented. Since we are dealing with the life-cycle of projects, the initial decision conditions may change over time, which could lead to the change of risks. Hence, a feedback and continuous assessment of risk through the entire life span of the project is very important in the process of whole life-cycle costing. This process should include tracking the effectiveness of the planned risk responses, reviewing any changes in priority of response management, monitoring the state of the risks, updating the whole life-cycle analysis accordingly and reviewing the economic performance indicators to check whether the investment decision is still valid or otherwise. In this way risk monitoring not only evaluates the performance of risk response strategies but also serves as a continuing feedback or audit mechanism.

The application of the above framework should take place during the early stages of asset development as well as at every project milestone, and should continue throughout the whole life of the asset. The information generated from the WLCC risk management framework should inform decision makers on which input data has the most impact on the WLCC result and how robust the final decisions are.

2.7 Summary

Evidence from research and practitioners alike has indicated strongly why WLCC has been treated with mistrust – the failure of models to adequately deal with uncertainty. Forecasts by their very nature can be risky to varying degrees, and stakeholders will imperil investment in capital projects if they are not fully equipped with the facts surrounding this uncertainty. It is therefore the responsibility of the analyst to ensure that the WLCC framework deals effectively with risk and provides the necessary information required to make effective decisions. In later chapters of this book we will look at the techniques and procedures that are available to the analyst to develop well-rounded WLCC models that are reliable and accurate.

References

Ashworth, A. & Hogg, K. (2000) *Added Value in Design and Construction*. Longman Publishing, London.

Bennett, J. & Ferry, D. (1987) Towards a simulated model of the total construction process. *Building Cost Modelling and Computers*. E. & F.N. Spon, London, pp. 377–85.

Bird, B. (1987) Costs-in-use: principles in the context of building procurement. *Construction Management and Economics*, Vol. 5 Special Issue.

Clift, M. & Bourke, K. (1998) *Study on Whole Life Costing for the Department of the Environment, Transport and the Regions (DETR)*. Report CR 366/98, Building Research Establishment, Watford.

Department of Industry (1977) *Life-cycle Costing in the Management of Assets*. The Stationery Office, London.

Egan, J. (1998) *Rethinking Construction*. The report of the Construction Task Force to the Deputy Prime Minister, on the scope for improving the quality and efficiency of UK construction. Department of Trade and Industry. The Stationery Office, London.

Ehlen, M.A. (1997) Life-cycle costs of new construction materials. *Journal of Infrastructure Systems*, December, 129–33.

Ferry, D.J. & Brandon, P.S. (1991) *Cost Planning of Buildings*, 6th edn. Blackwell Publishing, Oxford.

Flanagan, R. & Norman, G. (1983) *Risk Management and Construction*. Blackwell Science, Oxford.

Flanagan, R., Kendell, A., Norman, G. & Robinson, G.D. (1987) Life-cycle costing and risk management. *Construction Management and Economics*, **5**(4), S53–S71.

Fuller, S. & Petersen, S. (1996) *Life-cycle Costing Manual for the Federal Energy Management Program*. NIST Handbook 135. US Government Printing Office, Washington.

Kirk, S.J. & Dell'Isola, A.J. (1995) *Life-cycle Costing for Design Professionals*. McGraw-Hill, New York.

Koller, G. (1999) *Risk Assessment and Decision Making in Business and Industry: a practical guide*. CRC Press, New York.

Latham, M. (1994) *Constructing the Team*. Final report of the government/industry review of procurement and contractual arrangements in the UK construction industry. Department of the Environment. The Stationery Office, London.

Newton, S. (1991) An agenda for cost modelling research. *Construction Management and Economics*, **9**(2), 97–112.

Pilcher, R. (1992) *Principles of Construction Management*, 3rd edn. E. & F.N. Spon, London, p. 243.

Plenty, T., Chen, W. & McGeorge, W. (1999) Accrued value assessment – a dynamic approach for investment appraisal and facilities management. In *Proceedings of 8th International Conference on Durability of Building Materials and Components*, 30 May–3 June, Vancouver, Canada, pp. 1765–73.

White, K.H. (1991) Building performance and cost-in-use. *The Structural Engineer*, **69**(7), 148–51.

3 Key Decisions in the Whole Life-cycle Costing Process

3.1 Introduction

Clients or investors can view the WLCC process of a building facility as a sequence of investment decisions. The process contains parallel and interrelated phases. A conceptual model of such a process is shown in Fig. 3.1 and the model integrates six sequential steps:

(1) Justification for investment and client's requirements
(2) Conceptual development stage
(3) Design stage
(4) Production stage
(5) Operational stage
(6) End of economic life stage.

The horizontal axis denotes a project phase or option (portfolio of investment) and the critical time at which key decisions and whole life cost analysis need to be performed or updated. The vertical axis denotes the accumulative cost and the cost of acquiring an option within the life-cycle

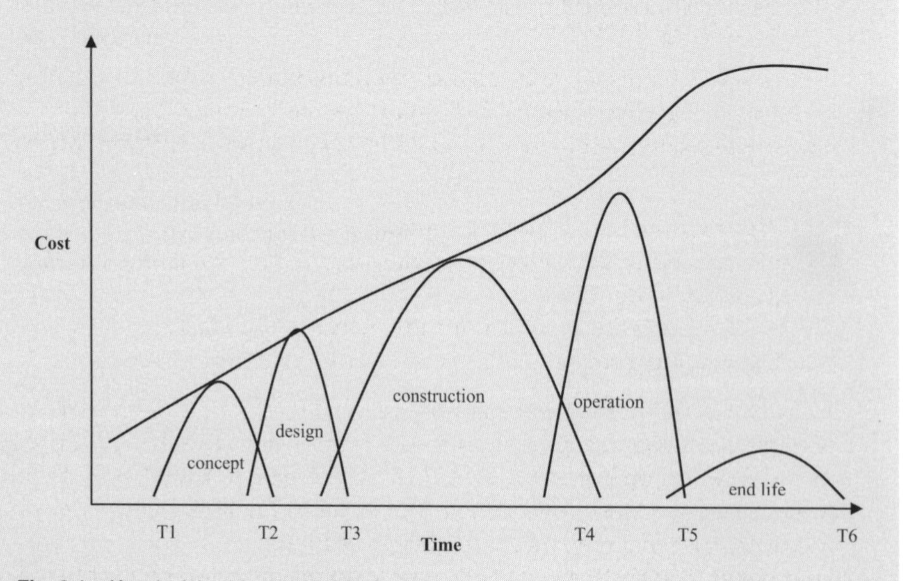

Fig. 3.1 Key decisions in the WLCC process.

of a project. This also can be viewed as the acquisition cost of moving from one stage to another within the life span of a project or facility. The application of such a model should take place as early as possible and continue until the end life of the asset. A sequential, in-depth discussion of the specific elements in this model is presented next.

3.2 Justification for investment and extraction of client requirements

This stage of the WLCC process is concerned with the development of a robust business case analysis. Objectives and assumptions about projects must be developed and scored according to their importance to the needs of investment. Figure 3.1 shows that the whole life-cycle cost of a project can be thought of as a set of sequential key decisions (stakeholder decisions) at different points in time: at time T = 0 a need for investment has to be justified and identified. The information required for computing whole life-cycle costing at this stage is minimal. The main decisions here are those related to why an investment is required in a particular facility and what are the requirements. Hence, the decision to invest should be based on the benefits that are perceived by the investors. These could include a combination of the following:

- Social
- Economical
- Strategic advantages over competitors
- Productivity improvements.

If the above benefits are to be realised then investment in building asset objectives have to be clearly defined. A report by the National Audit Office (2001) has identified four steps that need to be taken in making project objectives clear:

(1) Selecting the best projects to pursue
(2) Making clear the outcome of the project and what the investment is expected to deliver
(3) Determining the best form for the delivery of projects
(4) Developing a robust business case for investment.

All investments in the built environment facilities are also governed by a set of requirements as set by investors. The importance of ranking these requirements depends on many circumstances at the time of investment. It can also be said that investors have to take into account the needs of their customers and make sure that these needs are integrated into the overall business case for investment. The traditional client's needs usually are based on the three aspects of time, cost and quality performance criteria. But the built environment agenda has changed over the years, and all investment requirements need to be based on the whole life-cycle performance of facilities. Hence, the following needs are also essential:

- Sustainability
- Environmental requirements
- Ease of functional reconfiguration
- Impact on quality of life
- Time and cost requirements
- Operational strategies
- End of use strategies
- Quality requirements.

Apart from defining the objectives and needs for investment, clients are required to make the following cost decision assumptions:

- Land and demolition cost if required
- Life expectancy of investment or facility/economic life
- All legal and professional fees
- Physical characteristics of the facility, floor area, number of floors, etc.
- Development including all site costs
- Operation
- Maintenance and replacement
- Disposal
- Residual value
- Salvage value
- Status quo cost
- Cost of finance and economical factors like rates, taxes, etc.
- Income
- Budget and time limitations
- Building the team that represent and advise
- Limits on risk reserves
- Balance required between time, cost and quality
- Ability to deal with legal obligations
- Availability of financial and all required resources.

Perhaps one of the most important requirements for a successful whole life costing risk assessment is having a well-documented budget estimate to serve as a baseline. This is paramount at this stage because the WLCC of a facility will probably evolve as the investment goes through its logical development stages. That is why it is important to elicit and identify clients' cost priorities, and establish procedures and metrics for defining these costs. All assumptions and key decisions taken should be documented and updated as the project planning progresses through its life-cycle. In taking the above WLCC key decisions investors should not confuse assumptions with facts. An assumption is a hypothesis related to unknown costs as opposed to a fact and relates to a future occurrence; it involves a degree of uncertainty. For this reason, it is also essential that clients should set minimum and maximum limits on the above cost assumptions as low, most likely and high. The WLCC budget should represent the most likely costs, based on the available information.

The authors suggest that all key decisions and benefits identified at this stage should be ranked in order of importance, and the risk associated with them should be scored using an analytic hierarchy process. It is insufficient to present the decision maker with a set of alternatives whose costs and benefits are based on most likely factors and assumptions. The decision maker needs to be informed about how well the WLCC budget will hold up under changes to factors and assumptions. The outcome of the analysis carried out in this stage should inform the decision maker about the risks and viability of investing in the next phase of the development process (Flanagan & Tate 1997).

3.3 Key decisions at the conceptual development stage

As demonstrated in Fig. 3.1, at the time T = 1, after the need is identified and justified, investment in concept development is possible. Time T = 1 is equivalent to the RIBA outline plan of work, outline proposals and scheme design stages. The overall viability of investment and clients' needs and requirements has been established in time T = 0. Time T = 1 period is aimed at translating clients' needs and requirements into an overall sketch design, which incorporates the layout and construction methods of the facility. It is anticipated little costing information will be known about the client needs at this time, hence a basic estimate of costs and benefits for the desired facility and its operation should be acceptable. This can be based on rules of thumb, cost per square metre or other preferred methods.

All the benefits and WLCC information should be consistent with the overall client business case and customer needs. As a minimum costing requirement, costs and benefits should be identified through the whole life-cycle of the project and presented in present value. Benefits, financial and non-financial, should be quantified to the maximum extent possible. Based on WLCC analysis performed at this stage, including risks associated with the anticipated investment, the decision maker has the choice to:

- Invest in the next phase
- Delay investment, *or*
- Completely abandon the project.

In order to take these decisions the investor must base his/her actions on whole life-cycle costing analysis, economic performance indicators and the associated level of risk anticipated from the preferred course of action or decision. Also key decisions relating to different design forms and construction methods of the project need to be considered and alternative solutions evaluated using WLCC and environmental life-cycle assessment methods. All cost implications of the generated alternatives should be considered. Major decisions that affect the production and operation of facilities will also be considered here.

Bad decisions at this crucial stage could lead to lower quality and a large maintenance and operation budget and possible early disposal of the facility. WLCC decisions at this stage may involve:

- Substructure
- Structural framework
- Upper floors
- Environmental services
- Life expectancy of the major components
- Frequency of replacement and maintenance through the economic life the facility
- Internal finishes/partitions/doors/stairs/fittings
- External shell/walls/windows/doors
- Roofing systems
- External works
- Environmental impacts
- Operational
- Production
- Disposal.

Some form of WLCC analysis at this stage is required. This could be based on generic data from past or similar projects using rules of thumb, experience and historical data wherever possible. WLCC should be based on the baseline budget, and assumptions made at the justification stage. It is feasible, if the size, location and occupants of a facility are known approximately, to generate several assumptions relating to capital cost, operational costs and maintenance costs based on £/m². This information can then be fed into simulation software and tens of scenarios can be generated from these basic assumptions. Probably, at time T = 1, stakeholders are looking for direction of the cost movement rather than an exact calculation. Based on these scenarios, various cash flow profiles of the whole life-cycle of an asset should be projected.

3.4 Key decisions at the detailed design stage

Completion of phase 1 allows investment in period T = 2 to be started. The purpose of the detailed design stage is to produce a full design of every part and component of the facility in conjunction with other design stakeholders. Client requirements should be translated into functional requirements (functional design). The best design alternative should be selected based on risk, WLCC, and benefits (financial and non-financial). All outstanding decisions should be finalised, especially those relating to design specification, construction methods and associated costs. A systematic economic analysis is required for justifying the design options and their future consequences on operational and maintenance costs of the developed facility. Environmental life-cycle analysis is also essential at this stage due to its effect on the value and operation of the designed facility. Strict guidelines are required for selecting the best design alternative. These should be developed in conjunction with clients depending on their aspirations and requirements. Such guidelines could be based on the functionality of the facility and the economic and environmental performance indicators. There should be sufficient detail to fulfil adequate benefit analysis, and WLCC and budget estimates. Detailed consideration should be given to the likely effect of maintenance and operational costs associated with the selected design solution. This will result in design benefits analysis. In this analysis various costs scenarios of design decisions should be projected through to the end of the design's economic life. Decisions taken at this stage should include:

- Updating all decisions from the previous stages
- Final detailing specifications
- Construction detail production
- The impact of design solution on operation and maintenance costs
- Arrangements for production
- Comparing the costs and benefits of each option
- Detailed WLCC analysis for major components.

3.5 Key decisions at the production stage

Making the decision to invest in the production stage is critical. It has great financial consequences to clients. That is why WLCC at the end of the design phase must be sufficiently detailed and scrutinised from all aspects.

At period T = 4, as shown in Fig. 3.1, investment in production and development of a facility is possible. Ideally before a production decision is taken, an updated narrative review of the last phase should be conducted to make sure the initial parameters used in the economic analysis and selection processes have not changed. Discussion and analysis of the key decisions should be conducted to ascertain that WLCC parameters are not changed and the identified risk and risk responses strategies are still valid. Also it is important to re-examine that the designed facility still meets the need of clients and their customers. If there are new additions or changes of state for any reason (e.g. changes in technology, user requirements, regulation, etc.), these changes must be incorporated into WLCC analysis and risk response strategies and a new costs and benefits analysis should be produced. If the detailed design assessment or evaluation was proved viable, then a decision should be taken to expand resources for the development of the designed facility. WLCC at this stage requires only rigorous monitoring and updating. Stringent guidelines should be established to make sure that the project stays within the planned budget and quality and meets the original client specification and brief. Any changes to the project status that have an effect on the WLCC analysis should be identified and the economical benefits analysis must be updated to reflect changes. A better approach is that no decision should be taken regarding changes to project scope or specification until the WLCC analysis is updated, and economic performance measures ought to be used to assist in making these changes. The project risk profile also needs to be updated. Decisions taken at this stage should include:

- Updating all previous decisions as necessary
- Contractual arrangements
- Construction schedule
- Payment methods
- Reporting and monitoring mechanisms
- Construction technology/methods (Ashworth & Hogg 2000)
- Buildability aspects
- Site processes
- Resources, especially workmanship level and quality of materials
- Commissioning processes.

3.6 Decisions at the operational stage

At time T = 5, after completion of the production phase, investment in the operation phase is possible. The purpose of this stage is to conduct a post-occupation and operational assessment of the developed facility. The majority of the cost occurs at this stage so careful operational planning and optimisation is required. Plans for short-term and long-term operation are required. Maintenance and modernisation plans and budgets should be forecast based on the yearly performance of the facility in question. WLCC analysis should be updated to reflect the changes in physical and other resources requirements. This analysis should be conducted continuously (preferably annually) to determine if the existing facility continues to satisfy the original cost and functional objectives and to show if the facility requires modernisation or should be terminated. Also, a continuous risk assessment (including issues related to operational risk assessment) regarding operational costs and obsolescence of the facility, is a prerequisite at this stage. Several analysis techniques may be used singly and collaboratively to assist the decision maker in this stage, as discussed in Chapter 5. The outcome of this stage will inform decision makers when they should de-invest in the existing facility. Decisions taken at this stage should include:

- Strategies for reconfiguration/adaptation to accommodate changes in client needs
- Frequency and schedule for internal and external fabric repair, maintenance and replacement (Kirk & Dell'Isola 1995)
- Strategies for fit-out improvement
- Mechanisms for evaluating operational termination
- Frequency and schedule for mechanical and electrical (M&E) repair, maintenance and replacement
- Risk reserves for insurance, charges and taxes

- Budget and mechanism for security, water, sewerage, cleaning and waste disposal
- Mechanism for facilities and project management if applicable
- Frequency and schedule for grounds maintenance.

3.7 Decisions at the end of economic life stage

If at any time of the operation stage the updated WLCC analysis through the economic performance measures shows that the facility is no longer economically viable, the decision may be taken to invest in the disposal stage. At time T = 6, the disposal operation starts. Here owners are required to consider the environmental consequences of the disposal and try to maximise return from any salvage. Decisions taken at this stage should include:

- Dispose, retain or refurbish options
- Mechanism to recycle components
- Reserve for waste disposal
- Demolition methods
- Site and land clean-up implication
- Site resale or redevelopment options
- Risk reserves for end life.

3.8 Summary

The actual sequence of decisions described in Fig. 3.1 depends upon the particular project procurement route strategy. Some are sequential decisions, which could be attributed to the development process, whilst others may be associated with procurement and operational issues. This concept of depicting the whole life-cycle of a building as a series of investment decisions is generic and could include a variety of different development scenarios. We envisage the development of a building as a set of staged investments. Risk analysis and WLCC are conducted at the beginning and updated at the end of each stage in the project, and commitment is made to invest in the next stage of the project subject, of course, to the outcomes of the WLCC analysis. Whilst a large number of publications provide some discussion and examples of risk mitigation strategies at some of these key stages, these are primarily fragmented and do not deal with the risk throughout the entire whole life-cycle of the building. A systematic, theory-based approach to risk modelling is therefore lacking. We suggest that identifying, quantifying, responding and managing cost and risk at each investment stage of a project by fine tuning the portfolio at each stage is a useful framework for understanding cost and risk strategies, and the decisions made as a result throughout the whole life-cycle.

References

Ashworth, A. & Hogg, K. (2000) *Added Value in Design and Construction*. Longman, Harlow.

Flanagan, R. & Tate, B. (1997) *Cost Control in Building Design*. Blackwell Publishing, Oxford.

Kirk, S.J. & Dell'Isola, A.J. (1995) *Life-cycle Costing for Design Professionals*. McGraw-Hill, New York.

National Audit Office (2001) *Managing the Relationship to Secure a Successful Partnership in PFI Projects*. National Audit Office, London.

4 Fundamentals of Whole Life-cycle Cost Analysis

4.1 Introduction

The detailed nature of WLCC requires the analyst to have at their disposal a variety of mathematical and analytical skills to perform the task effectively. The accessibility and quality of the data available and skills of the analyst will normally dictate the mathematical techniques required to perform a WLCC analysis. This chapter provides an insight into some of the mathematical fundamentals in WLCC. The chapter begins by examining the fundamentals of modelling WLCC and introduces time value of money concepts. Next the application of discounting techniques and economic performance measures in WLCC modelling is introduced. The chapter then moves on to consider forecasting costs in WLCC, critically assessing the use of regression, time series, utility theory and artificial intelligence models in WLCC forecasting. Finally, the chapter explains how and why benchmarking and performance indicators are essential for WLCC decision making processes.

4.2 Concepts of the time value of money

The time value of money concept is a reflection on the fact that present capital (cash in hand) is more valuable than a similar amount of money received in the future. If this assumption is accepted as fact, then benefits and costs are worth more if they are achieved much earlier. Time value of money computation is based on present value and discounting techniques. The present value method captures the time value of money by adjusting through compounding and discounting cash flows to reflect the increased value of money when invested.

4.2.1 *The discounting process*

Discounting is a process for calculating the amount today that a sum of money in the future is worth using a specified discounting rate. Thus, discounting translates projected building whole life costs into present value. The challenge here is how one can specify a discount rate. This is due to the uncertainty on how to project future changes in the discount rate.

Since the discount rate reflects the future value of money concept, it can be a real discount rate adjusted to eliminate the effects of inflation or a nominal discount rate adjusted to reflect expected inflation. Selection of the correct discount rate depends upon whether the benefits and costs are measured in real or nominal terms. A real discount rate that has been adjusted to eliminate the effect of expected inflation should be used to discount constant whole life-cycle costs. Subtracting expected inflation from a nominal interest rate could approximate a real discount rate. A nominal discount rate that reflects expected inflation should be used to discount nominal whole life-cycle costs. Market interest rates are nominal interest rates in this sense.

4.2.2 The compounding process

Compounding is in essence the value of WLCC at a specified date in the future that is equivalent in value to a specified sum today. The compounding process is computed according to the following formula:

$$PV = S\left(1 + \frac{r}{100}\right)^{-t}$$

Where:
PV = present value
S = the future value of capital
r = discount rate
t = time or period of analysis.

4.2.3 Net present value

Net present value (NPV) is an approach used in WLCC budgeting where the present value of cash inflow is subtracted from the present value of cash outflows. Thus, a building asset's net contribution to the client business is the difference between the discounted present value of benefits and the discounted present value of costs. In this sense NPV is a metric for measuring the net value of an investment in building assets in today's money. Benefits can be monetary or any other aspects as long as monetary values are assigned to benefits and costs. NPV is calculated using the following formula:

$$NPV = \sum[PV(b) - PV(c)]$$

Where:
$PV(b)$ = discounted present value of benefits
$PV(c)$ = discounted present value of costs.

4.3 WLCC calculation models

As discussed in previous chapters, WLCC enables asset stakeholders to forecast the cost of acquiring, owning, operating, maintaining, reconfiguring and disposing of an asset. This concept can be applied to the whole building or part/components of a building, such as mechanical services. The basic deterministic approach to forecasting WLCC is very simple, requiring only a study period, a discount rate and yearly cost forecast estimates for competing alternatives. There are three possible ways for modelling WLCC computation.

4.3.1 Deterministic calculation of WLCC

In the deterministic approach, the analyst computes the present value of an investment from the time series of projected cash flows using the discount rates specified. In the method used to estimate NPV, all future costs and benefits must be discounted. The discount factors can be assumed to be real or nominal, depending on the requirements of the stakeholders. Here, all WLCC terms are computed using a single value. The results can then be subjected to sensitivity analysis to determine WLCC variation in case of changes to initial input parameters. The following is a generic formula for the WLCC present-value model, which is derived from the formula to calculate NPV:

$$WLCC = C_P + \sum_{t=0}^{n} \frac{C_t}{(1 + d)^t}$$

Where:

$WLCC$ = total WLCC in present value of total ownership of a building asset

C_t = sum of relevant whole life costs, including initial capital costs and future costs up to the end life of the asset or period of study, less any positive cash flows, such as residual value of the asset or land resale value

n = number of years of the period study, usually the service life of the asset or components

d = discount rate that captures the time value of money by adjusting cash flows to the present

C_P = initial capital costs.

The assumption in the above WLCC computation states that all costs be identified by year and by amount with certainty (i.e. there is no probabilistic nature in the values).

4.3.2 Stochastic calculation of WLCC

From a purely financial perspective, all or some WLCC cost centres can be modelled probabilistically. Each WLCC term in the above equation can be assumed to be a probability distribution rather than a deterministic value. These distributions can be derived from the expected value and variance of WLCC estimates under different financial scenarios over the study period. The model is based on the assumption that the WLCC cost centres, discount rate and the study period are randomly distributed according to one of the theoretical probability distribution forms that will be discussed in Chapter 5. This requires that each cost centre element be treated stochastically and the cash flow (PV) in each year of the study period is described as probability density functions or uncertain cash flow profiles. These

PV cash profiles are considered to be variable from one year to another depending on inflation and interests rates, etc. Hence if the probability distribution function (PDF) or cash flow profile of each WLC cost centre and discounting parameter is known or can be simulated, the total WLCC of ownership cost can be simulated using the following equation:

$$f(PV) = f(C_P) + \sum_{t=0}^{n} \frac{f(C_{ti})}{(1 + f(d))^t}$$

Where:

$f(PV)$ = probability distribution function of total WLCC in present value of total ownership of a building asset

$f(C_{ti})$ = probability distribution for whole life cost centre i in period t, including initial capital costs and future costs up to the end life of the asset or period of study

n = number of years of the period study, usually the service life of the asset or components

$f(d)$ = probability distribution of discount rate used to adjust cash flows to present value

$f(C_P)$ = probability distribution of initial capital cost.

In the above model, probability distributions are assigned to the input variables and a probability distribution is derived. The output probability (in our case WLCC) is defined as a risk profile, showing the likelihood of the different possible net present values. The central limit theorem states that the sum of independently distributed random variables tends to be normally distributed as the number of terms in the above equation summation increases. Since most periods of study are over 25 years, one expected $f(PV)$ to be normally distributed with the following mean and variance values:

$$\sigma(PV) = \sum_{t=0}^{n} \frac{\sigma(C_{ti})}{(1 + d)^t}$$

$$\sigma^2(PV) = V(PV) = \sum_{t=0}^{n} \frac{V(C_{ti})}{(1 + d)^{-2t}}$$

These two equations are only valid if WLCC cost centres (C_{ti}) are statistically independent. If this is not the case the computation should be modified to take into consideration the correlation between WLCC cost centres.

4.3.3 Fuzzy calculation of WLCC

WLCC computations based on probability theory can generate powerful information if the assumptions about asset future circumstances are right. However, it is not always the case that uncertainty associated with WLCC analysis fits the probability theory assumptions. Fuzzy set theory is an important tool for modelling uncertainty or imprecision arising from human perception. Human judgement is involved in all aspects of WLCC. Therefore, a rational approach towards WLCC modelling is to take into account human and processes subjectivity. WLCC and present value parameters are usually estimated by using expert judgement and statistical techniques. Present value computation based on fuzzy numbers can capture the difficulties in estimating the attributes of PV and WLCC. For example, the following is a formula for computing fuzzy PV (Kahraman *et al.* 2002).

$$P\widetilde{V} = \left(\sum_{t=0}^{n} \left(\frac{\max(P_t^{1(y)}, 0)}{\prod_{t'=0}^{t} (1 + t_{t'}^{r(y)})} + \frac{\min(P_t^{1(y)}, 0)}{\prod_{t'=0}^{t} (1 + r_{t'}^{l(y)})} \right), \right.$$

$$\left. \sum_{t=0}^{n} \left(\frac{\max(P_t^{r(y)}, 0)}{\prod_{t'=0}^{t} (1 + r_{t'}^{1(y)})} + \frac{\min(P_t^{r(y)}, 0)}{\prod_{t'=0}^{t} (1 + r_{t'}^{r(y)})} \right) \right)$$

Where:
$P_t^{1(y)}$ = left membership representation of the whole life-cycle cost at time t
$P_t^{r(y)}$ = right membership representation of the whole life-cycle cost at time t
$r_t^{1(y)}$ = left membership representation of the discount rate at time t
$r_t^{r(y)}$ = right membership representation of the discount rate at time t.

Similarly, there are formulae for the analysis of fuzzy future value, fuzzy benefit-cost ratio, fuzzy payback, etc. For more detail on the fuzzy set theory approach to capital budgeting, readers are referred to Kahraman *et al.* (2002) and Abdelkader & Dugdale (2001) and this is covered in more detail in Chapter 5 of this book.

4.4 Measuring economic performance in whole life-cycle costing

When using WLCC analysis techniques, it is simply inadequate to forecast a cost value without providing the analyst or stakeholder with the ability to draw inference from the results. In previous applications of WLCC, where the concept has been used solely as a competing options decision making tool in construction projects, the results simply inform the analyst which project is more economically viable. However, in other applications this information is irrelevant, as projects are not being compared.

Using economic performance indicators in conjunction with a WLCC analysis helps to provide the information that is required, on cost performance of a building over a specified time period. A variety of techniques are available for this purpose. This section explores the use of benchmarking and key performance indicators and assesses the viability of including such approaches in a WLCC framework.

The necessity for a range of economic performance measures in cost analysis is required for WLCC control. WLCC control is the process whereby a stakeholder ensures that he/she is pursuing strategies and actions which will enable him/her to achieve WLCC goals. The measurement and evaluation of performance are central to control and this is addressed by means of the following questions (Foundation for Performance Measurement 1999).

- What has happened?
- Why has it happened?
- Is it going to continue?
- What are we going to do about it?

The first question can be answered by performance measurement. The WLCC analyst will then have to hand far more useful information than would otherwise be the case in order to answer the other three questions. By finding out what has actually been and is happening, the WLCC analyst can determine with considerable certainty the direction in which the investment is going and, if all is going well, continue with the status quo. Or, if the performance measurements indicate that there are difficulties on the horizon, decision makers can take steps to remedy the situation.

In WLCC, performance measurement can play an important role in giving this vital whole life information from the results of the analysis. The next section looks closely at some of the techniques available to elicit this information.

4.4.1 Simple payback (SPB) and discounted payback (DPB)

SPB and DPB are the most basic measures of the amount of time it takes to recover the initial investment in capital. Both are expressed as the period that has elapsed between the beginning of the service life period and the time at which cumulative savings are just sufficient to cover the initial capital cost of the investment decision (WBDG 1993). In terms of WLCC (competing alternative decision making) the problem of both measures is that they are not valid for comparing multiple, mutually exclusive project alternatives, nor for ranking alternatives. However, they are best employed as screening methods for projects that are so clearly economical that a fuller WLCC treatment of the project is uneconomical and unwarranted. The general formula for calculating payback is given by:

$$\sum_{t=1}^{y} \frac{(S_t - \Delta I_t)}{(1 + d)^t} \geq \Delta I_0$$

Where:

y = minimum length of time over which future net cash flows have to be accumulated in order to offset initial capital cost investment

S_t = savings in operational costs in year t associated with an alternative project

ΔI_0 = initial investment costs associated with the alternative

ΔI_t = additional investment related costs in year t, other than investment costs

d = discount rate.

4.4.2 Return on capital employed (ROCE)

ROCE, sometimes referred to as return on net assets (RONA), is probably the most popular ratio tool for measuring general economic performance in relation to the capital invested in the building (Van Horne 1995). ROCE defines capital invested in the building as total assets less current liabilities, unlike 'return on total assets' (ROTA), which measures profitability in relation to total assets.

$$ROCE(\%) = \frac{\sum P_n}{\sum TC_E} \times 100$$

Where:
$\sum P_n$ = net profit before income and taxes (NPIT)
$\sum TC_E$ = total capital employed (CE).

Capital employed may be defined in a variety of ways, the most common being fixed assets plus working capital, i.e. current assets less current liabilities. This definition reflects the investment required to enable a business to function.

4.4.3 Net savings

Net savings (NS) measure is a variation of the net benefits (NB) measure for the economic performance of a building. The NB method assesses the difference between present day benefits (in money terms) and present day costs over the period of the WLCC analysis period. NB is primarily used to justify investment decision making by highlighting the positive cash flows in building expenditure.

The net savings measure finds its use in assessing the benefits that are expected to occur primarily in the form of future operational cost reductions such as energy consumption. The NS method calculates the net amount, in present value pounds, which an investment decision is expected to save over the study period (or economic service life). As NS is expressed in PV terms, it represents savings over and above the amount that would be returned from investing the funds at the minimum expected rate of return (i.e. the discount rate). The NS for a project, relative to a designated base case, is calculated by simply subtracting the WLCC of the alternative project under consideration from that of the base case such that:

$$NS = WLCC_{BaseCase} - WLCC_{Alternative}$$

It is considered that as long as the NS value is greater than zero, then the project under consideration is economically cost effective relative to the base case. Naturally, a relationship exists between NS and WLCC. In the evaluation of multiple, mutually exclusive project alternatives, the project with the highest

NS will also have the lowest WLCC, so both methods are entirely consistent and interchangeable. The principle difference though between WLCC and NS is that in the former there is no requirement for a base case to evaluate the costs.

The use of NS can also be extended to individual cost differences between the base case and the alternative. Examples such as capital costs, maintenance cost, etc. can benefit from this. However this does require additional calculations as opposed to the simple method above, but it is useful as this value is needed in the calculation of other methods such as savings to investment ratio (SIR) and adjusted internal rate of return (AIRR). Calculating NS using individual cost differences is useful as a check to ensure that SIR and AIRR calculations are based on correct intermediate calculations. That is, the NS should be exactly the same whether computed by the comparison of WLCCs or by using individual cost differences. For the latter, the following equations can be employed in the calculation of NS (operational costs, i.e. energy):

$$NS_{A:BC} = \sum_{t=0}^{N} \frac{S_t}{(1+d)^t} - \sum \frac{\Delta I_t}{(1+d)^t}$$

Where:

$NS_{A:BC}$ = NS, in present value, of the alternative (A), relative to the base case

S_t = savings in year t in operational costs associated with the alternative

ΔI_t = additional investment related costs in year t associated with the alternative

t = year where base date $t = 0$

d = discount rate

N = number of years in study period (service life).

4.4.4 Savings to investment ratio (SIR)

The savings to investment ratio is a measure of economic performance for a project that expresses the relationship between its savings and its increased investment cost (in present value terms) as a ratio. (WBDG 1993) SIR is particular useful to WLCC where, in decision making, increases in initial capital expenditure result in lower long-term operational cost savings. It is a hybrid of the conventional benefits to cost analysis. As SIR is a measure of economic performance like NS, it requires a base case to compare.

As a general rule of thumb, it is considered that an SIR return greater than 1.0 is suitable for economic justification of a project alternative (i.e. if savings are greater than its incremental investment costs, and its NS is greater than zero). The difference though in terms of a comparison with NS is that the lowest WLCC for a project is not necessarily the project with the highest SIR. SIR should not really be used for choosing among projects but should be used as a ranking tool for comparing other independent projects for allocating investment funding.

The general rule for calculating the SIR of operational costs, for example, is given by:

$$SIR_{A:BC} = \frac{\sum_{t=0}^{N} S_t \div (1 = d)^t}{\sum_{t=0}^{N} \Delta I_t \div (1 = d)^t}$$

Where:

$SIR_{A:BC}$ = ratio of present value savings to additional present value investment costs of the alternative, relative to the base case

S_t = savings in year t in operational costs attributable to the alternative

ΔI_t = additional investment related costs in year t attributable to the alternative

t = year of occurrence (where zero is base date)

d = discount rate

N = length of study period (service life).

4.4.5 Internal rate of return

The internal rate of return measurement (IRR) is disputed by many as simply a theoretical arithmetic result as opposed to an economic measure of performance. The IRR of a project can be defined as the rate of discount which, when applied to the project's cash flows, produces a zero NPV, so in general terms the IRR is the value for r which satisfies the expression:

$$\sum_{t=0}^{N} \frac{A}{(1 + d)^t} = 0$$

The decision rule for IRR is that only projects with an IRR greater than or equal to a predefined cut-off point should be accepted. This cut-off rate is usually the market rate of interest (i.e. the discount rate that would have been used if an NPV analysis were undertaken instead). All other project investment opportunities should be rejected. The logic behind IRR is similar to that of NPV. The market interest rate reflects the opportunity cost of the capital involved. Thus, to be acceptable, a project must generate a return at least equal to the return available elsewhere on the capital market.

4.4.6 Adjusted internal rate of return (AIRR)

The adjusted internal rate of return (AIRR) is a measure of the annual percentage yield from a project investment over the service life of the building (or study period in basic terms). Like the NS and SIR measures, it is a relative calculation that needs a base case for comparison.

AIRR is used to compare against the minimum acceptable rate of return (MARR). MARR is defined as (Pouliqnen 1970):

> 'The smallest amount of revenue considered acceptable for an organisation to undertake a project. Typically, MARR is equal to the cost of capital plus a return. Sometimes referred to as the hurdle rate.'

This is generally equal to the discount rate used in the WLCC analysis. If the AIRR is greater than the MARR, then the project can be defined as economic; if less it is deemed unworthy of investment. If AIRR is equal to the discount rate, this is break even and hence economically neutral. AIRR can be used in the same way as SIR.

The AIRR, in contrast to the IRR measure, explicitly assumes that the savings generated by the investment decisions can be reinvested at the discount rate for the remainder of the service life. If these savings could be reinvested at a higher rate than the discount rate, then the discount rate would not represent the opportunity cost of capital. IRR implicitly assumes that interim savings can be reinvested at the calculated rate of return on the project, an assumption that leads to overestimation of the project's yield if the calculated rate of return is higher than the reinvestment rate. AIRR and IRR are only the same if the investment yields a single, lump sum payment at the end of the service life, or in the unlikely case that the reinvestment rate is the same as the IRR.

As discussed earlier, some dispute IRR as a performance measure in that more than one rate of return may make the value of the savings and investment streams equal, as required by the definition of the internal rate of return. This may be the case when capital investment costs are incurred during later years, giving rise to negative cash flows in some years. The formula for calculating AIRR is given by:

$$\frac{\sum_{t=0}^{N} S_t (1 + r)^{N-t}}{(1 + i)^N} - \sum_{t=0}^{N} \frac{\Delta I_t}{(1 + r)^t} = 0$$

Where:

S_t = annual savings generated by the project, reinvested at the reinvestment rate

r = rate at which available savings can be reinvested, usually equal to the MARR (i.e. the discount rate)

$\Delta I_t \div (1 + r)^t$ = present value investment costs on which return is to be maximised

4.4.7 Net terminal value

In the same way that a project can be valued in terms of today's cost, cash flow can equally be valued in terms of what it will be worth at the end of the whole life using the net terminal value (NTV) formula. The formula for NTV is directly analogous to that for NPV, where

$$NTV = A_1 (1 + r)^{n-1} + A_2(1 = r)^{n-2} + \cdots + A_n - C(1 = r)^n$$

The difference between NTV and NPV is that the former cumulates forwards to make a valuation at the end of the whole life, whereas the latter discounts backwards to a present day valuation.

4.4.8 Sinking funds

An important aspect of capital investment is the long-term cost associated with running and maintaining the building. We know that a significant proportion of WLCC is associated with the running and maintenance costs, but how can measures be taken to offset these, or at least plan for them in the future? The facility owner can set aside regular payments, based upon future forecasted running costs, so that these costs can be met, at least partially anyway. These regular payments are called sinking funds. Of course, future sinking funds will accumulate interest at the specified amount and the sinking fund calculation can be used to show how this additional investment F accumulates after n period of time, assuming a given initial investment of A_0 and interest rate i:

$$A_n = A_0 \times (1 + i)^n + \frac{F \times (1 + i)^n - F}{i}$$

An alternative view of a sinking fund is that it involves calculating the amount which, when invested annually at the marginal rate, will yield by the end of the service life a sum equal to the initial investment. However, in the case of multimillion pound complexes like state-run hospitals, sinking funds are more likely to be used for maintenance and replacement rather than recouping the full initial capital investment.

4.4.9 The benefit/cost ratio (BCR)

This ratio, which can take several different forms, is the ratio between the discounted sum of the benefits and the costs of a project. In its simplest form it may be expressed as:

$$BCR = \frac{\sum_{i=0}^{i=n} C_i (1 + r)^{-i}}{1}$$

This formula relates the net discounted benefits of the project to the investment necessary to initiate the project. Under conditions of certainty, where r is the organisation's cost of capital and the project can be classified as normal (that is, after the initial outlay, the following periods would be net benefits such that the BCR would exceed unity). In appraisal processes where risk is not explicitly considered, netting of benefits may give misleading results and in such circumstances it is therefore preferable to use a ratio which compares total benefits to total costs, such that:

$$BCR = \frac{\sum_{i=0}^{i=n} R_i(1 + r)^{-i}}{\sum D_i(1 + r)^{-i}}$$

Where R represents the cash receipts and D the cash disbursements of a period.

4.4.10 Total annual capital charge

The total annual capital charge (TACC) is found simply by adding together the depreciation charge and interest cost. Using N as the symbol to denote this total annual capital charge, then:

$$N = \frac{Cr(1 + r)^n}{(1 + r)^n - 1}$$

It should be noted that in the application of this formula, it is assumed that the whole of the initial capital C is depreciated. Where any of the initial capital is recovered at the end of a project's life (such as working capital, land, residual value), then the true cost is found by using the above formula for the depreciable capital, and adding to it the annual interest charge for the non-depreciable element. Thus where the total capital C can be split into a depreciable element C_d and a non-depreciable element C_n, the total annual cost is expressed in:

$$N = \frac{C_d r(1 + r)^n}{(1 + r)^n - 1} + C_n r$$

The above conventional investment appraisal techniques, which focus on cash flows represented by the costs and expected returns of a project discounted to a common base period, do not reflect the total value of capital expenditure choices which include intangible and non-monetary benefits as well as reduction of future costs and financial returns (Plently *et al.* 1999). That is why these economic measures must be supported by risk assessment analysis. It is important that the WLCC analysis evaluates and allocates identifiable value from capital and continuing costs to relevant stakeholders in the life-cycle of a project. This will allow the consideration of different stakeholders' objectives in the assessment of the life-cycle of facilities.

4.5 WLCC forecasting methods

The ultimate test of any forecast is whether or not it is capable of predicting future events accurately. Planners and decision makers have a wide choice of forecasting techniques, ranging from purely intuitive or judgemental processes to highly structured and complex quantitative methods (Makridakis & Hibon 1979). Forecasting is a significant part of the whole life-cycle cost exercise, and therefore the approaches to selecting the correct method should be fully investigated. At this juncture, it is pertinent also to distinguish between what constitutes a WLCC model and what constitutes a WLCC forecasting method. A WLCC model is an equation (such as a regression model) or set of equations representing the stochastic nature of the problem to be forecast. A WLCC forecasting method is the combination of an estimating procedure and a model (Mead 2000). In establishing a WLCC forecasting procedure, it is vital to ascertain the objectives or assumptions of the exercise (i.e. is the model dealing with seasonal or non-seasonal data, monthly or annual data?). Not only must this be established clearly, but also a well-rounded knowledge of the background information needed to formulate the problem must be established. Chatfield (1997) discusses the importance of this iterative process, the need to appreciate not only the theoretical aspects of selecting a model, but also the strategy that needs to be established in the modelling process.

4.5.1 WLCC modelling using regression techniques

Regression analysis enables the analyst to ascertain and utilise a relationship between a variable of interest, called the dependent variable or response variable, and one or more independent variables known also as predictor variables (Montogomery & Runger 1994). Regression models are primarily developed to interpolate information about the dependent variable from the knowledge contained within the independent variables (Wanous 2000). They can also be used to investigate the relationship between sets of WLCC variables.

Regression analysis is classified into linear and non-linear forms (Makridakis *et al.* 1993).

Multiple-linear models have been widely reported on in the literature: Kirkham *et al.* (1999) used a variety of multiple-linear regression models to forecast energy costs in local authority sport centres. However, further work by these authors (Boussabaine *et al.* 1999b) revealed that other techniques, notably fuzzy models and artificial neural networks, yielded more accurate models. Myers (1971) used multiple linear regression models to forecast electricity consumption in the UK using growth in retail sales and average temperatures as the independent variables. Local electricity suppliers used these models in practical application.

The main disadvantage of the linear regression technique is that it is unable to account for the non-linearity that might exist in the relationship between the dependent and independent variables. Non-linear regression is an attempt to address the problem, but it is highly dependent on user intervention. The equations are developed and entered manually and this can introduce a heavy bias. However, the ease of development and use of regression models makes them a popular method of forecasting. The application of both multiple linear and non-linear regression techniques in WLCC modelling might include:

- Modelling operational costs
- Modelling relationships between WLCC attributes
- Modelling maintenance costs.

4.5.2 WLCC modelling using time series techniques

A time series is an ordered sequence of values of a variable observed at equally spaced time intervals. Therefore, a time series model is a function that relates the values of a time series to previous values of that time series, its errors and other related time series. Time series forecasting methods are generally a more accurate approach to forecasting than regression models because many TS models can represent seasonality in data. A plethora of time-series forecasting methods currently exist and these were reported on in the M-Competitions: 1001 time series were used from the 1982 M-Competition (Makridakis *et al.* 1982); 29 series from the 1993 M2-Competition; and 3003 series from the 1998 M3-Competition. Mead (2000) in his work on the selection of forecasting methods identified the methods within three distinct groups: the naïve methods, exponential smoothing methods and the auto-regressive moving average methods (ARMA). Naïve methods assume a minimal time-series structure and comprise three methods.

The first is a long run average where an average of previous observations is used as a forecast.

The second group of methods is those that use exponential smoothing. Exponential smoothing methods are discussed by Makridakis *et al.* (1993), who identify six variations of exponential smoothing: linear, linear (Holt's two parameter method), Pegels' classification, quadratic (Brown's one-parameter method), seasonal and single. These six methods can be further subdivided into seasonal and non-seasonal methods. With respect to seasonality methods, the Holt-Winters multiplicative and Holt-Winters additive methods were reported on in Chatfield (1978), as techniques that were able to better represent trend and seasonality in data, such as that which would be expected of electricity consumption data. The final group, the ARMA methods, are sophisticated techniques and are explained in detail in Fildes *et al.* (1998), Box & Jenkins (1970) and Makridakis *et al.* (1993).

The use of time series in WLCC might include:

- Modelling energy consumption
- Develop operational cost indices
- Finding trends in operational costs.

4.5.3 WLCC modelling using artificial intelligence and fuzzy theory

Expert systems, fuzzy logic and artificial neural network (ANN) provide a powerful tool that can help WLCC analysts in modelling maintenance and running costs. Expert systems are generally good in representing knowledge and rules of thumb. Fuzzy techniques have the ability to exploit the tolerance for imprecision and uncertainty and represent the real world, whereas ANN has the ability to learn and generalise from examples, to produce meaningful solutions over time to compensate for changing circumstances, to process information rapidly, and to transfer readily between computing systems. These important properties of expert systems, fuzzy techniques and ANN are invaluable in the process of modelling WLCC decisions. The role of using artificial intelligence (AI) in the process of WLCC is supportive rather than computational. AI techniques can be used in the analysis of whole life decision attributes. For example:

- Forecasting maintenance and running costs
- Forecasting user demand (volume)
- Ranking WLCC attributes
- Establishing non-linear relationship between whole cost centres and contributing factors
- Analysing WLCC risks.

4.5.4 Selection of an appropriate WLCC forecasting method

It is very difficult to specify a rationale for selecting an appropriate forecasting technique. This is evident in Collopy and Armstrong (1992) who present 99 different rules to facilitate the selection of annual data from four forecasting methods. For data which exhibits trend and seasonality such as electricity cost data, a simple seasonal forecasting technique is best applied. Indeed, in Fildes and Makridakis (1995) evidence is presented of the out-performance of complicated forecasting techniques by simple seasonal methods. However, it is clear that some kind of integration between statistical measurement and expert inference is preferable, particularly in Armstrong and Collopy (1998) who concluded that the most reliable forecasting techniques were identified as the ones that used a combination of statistical testing and expert domain knowledge. The researchers also concluded that time-series data which exhibited trend and seasonality was more likely to yield an accurate forecast than other data. This would validate the discussion in Chatfield (1997), which recommends the use of the Holt-Winters forecasting technique for forecasting seasonal data.

However, is forecast accuracy the sole decision criterion in selecting the forecasting model? Studies by Yokum and Armstrong (1995) and Makridakis and Hibon (1979) challenged the theory that accuracy was not the only factor in selecting a forecasting model. Interpretation, cost/time and ease of use all figured highly in decision criteria as well. Software packages that are currently available provide the analyst with the ability to use several different techniques simultaneously so that the relative accuracies can be compared statistically. Kirkham *et al.* (1999) and Boussabaine *et al.* (1999a, 1999b) in their work on forecasting energy costs in sport centres advocated the use of Theil's U statistic as an appropriate form of differentiating between different methods. The software packages available tend to rank forecasting methods based upon the mean absolute percentage error (MAPE), as was the case in the M-Competitions, although a variation on this, the unbiased absolute percentage error was proposed in Collopy and Armstrong (2000). It is widely acknowledged however that if the analyst is making an a priori assumption as to which forecasting technique is best suited to the task, then the analyst's skill and intuition will be significant in terms of the success of the forecasting method.

4.6 Benchmarking and key performance indicators

This section aims to provide an awareness of the importance of these techniques to WLCC. These performance measures generally apply to WLCC assessment on the micro-scale. If WLCC methodologies are to be used beyond this level to draw inference from the results (i.e. to compare cost performance between other types of buildings), then benchmarking can be used as a means to achieve this.

4.6.1 WLCC benchmarking

Benchmarking generally is about comparing and measuring performance against that of others in many aspects of business. It is now regarded as a vital tool in measuring variables such as cost performance. Benchmarking involves answering two questions – what is better and why is it better? – with the aim of using this information to make changes that will lead to economic and managerial improvements. The best performance achieved in practice is the benchmark. A benchmark is 'the best in class' level of performance achieved for a specific cost centre or decision. It is used as a reference point for comparison in the benchmarking process. For example, benchmarking metrics are discussed in Massheder and Finch (1998), where the use of benchmarks is applied to UK facilities management systems. Benchmarks are well established in local authority auditing, performance measurement in production of mechanical and electrical components as well as administrative tasks. Benchmarks are usually calculated through the statistical analysis of key performance indicators. A good understanding of the problem under consideration, and data analysis and manipulation are essential. If incorrectly applied, benchmarks can provide misleading and sometimes incorrect information (CBPP 2001). WLCC benchmarks can be set up in the following areas:

- Ratio of capital to maintenance and operation (M&O) costs
- Components service life
- Disposal costs
- WLCC risk management reserves
- Waste management
- Utilities
- Cleaning
- Repairs and maintenance.

4.7 WLCC key performance indicators

A key performance indicator (KPI) is the measure of performance associated with a WLC cost centre on the macro-scale. The information provided by a KPI can be used to determine how the costs of running a building compare with the benchmark, and therefore can form a key component in an organisation's move towards best practice and value for money.

WLCC KPIs can take a variety of forms, which reflect the stakeholder interest. KPIs form part of many approaches to good business management and practice but it is important to note that KPIs should not just be viewed as an ad hoc or short-term measure. The nature of KPIs means that they should continually be reviewed and the information gained from them used effectively to increase productivity and economic efficiency. This therefore implies the possible use of KPIs in WLCC. The definition of WLCC described in Chapter 1 states that it should be a dynamic process. Therefore the methods to draw inference from the results should be dynamic too. The use therefore of KPIs is a dynamic approach and thus a framework should exist where these KPIs are regularly reviewed and adjusted if necessary. WLCC KPIs might include:

1. Service life of building components
2. Building components' eco-costs
3. Asset operation eco-costs
4. Disposal eco-costs
5. Components' deterioration rate
6. Fabric maintenance
7. Services
8. Maintenance
9. Overheads
10. Utilities
11. Cleaning
12. Percentage of current replacement value
13. Ratio of maintenance to capital cost
14. Ratio of operation to capital cost.

4.8 Summary

This chapter has provided a brief review of the principal mathematical techniques that have been used in LCC and WLCC techniques. This review is useful for understanding the principles of mathematical techniques in WLCC procedures. Also it provides a guide for the selection of the most suitable techniques in a WLCC exercise. The criteria used to select the appropriate model can vary significantly between scenarios. It is hoped therefore that this chapter has highlighted the importance of this. This chapter has also aimed to give a concise review of the main themes in performance measurement and its application to WLCC methods. It is clear that in using WLCC in the analysis of existing buildings, performance measurement should form an important part of the modelling process. This can enable stakeholders to draw inference from the results and make informed managerial decisions based upon these.

References

Abdelkader, M. & Dugdale, D. (2001) Evaluating investment in advanced manufacturing technology: a fuzzy set theory approach. *British Accounting Review*, **33**, 455–89.

Armstrong, J.S. & Collopy, F. (1998) Integration of statistical methods and judgement for time series forecasting: principles from empirical research. In *Forecasting and Judgement* (eds G. Wright & P. Goodwin). John Wiley and Sons, New York.

Boussabaine, A.H., Kirkham, R.J. & Grew, R.J. (1999a) Modelling total energy cost of sport centres. *Facilities Journal*, **17**, December, pp. 452–61.

Boussabaine, A.H., Kirkham, R.J. & Grew, R.J. (1999b) Estimating the cost of energy usage in sport centres: a comparative modelling approach. *Proceedings of the 15th Annual Conference of the Association of Researchers in Construction Management*. John Moores University, Liverpool, pp. 481–8.

Box, G.E.P. & Jenkins, G.M. (1970) *Time Series Analysis, Forecasting and Control*. Holden Day, San Francisco.

CBPP (2001) *Factsheet on Whole Life Costing*. http://www.cbpp.org.uk/resourcecentre/publications/document.jsp?documentID=115776. The Construction Best Practice Programme, Watford.

Chatfield, C. (1978) The Holt-Winters forecasting procedure. *Applied Statistics*, **27**, 264–79.

Chatfield, C. (1997) Forecasting in the 1990s. *Journal of the Royal Statistical Society*, **46**, 461–74.

Collopy, F. & Armstrong, J.S. (1992) Rule-based forecasting: development and validation of an expert systems approach to combining time series extrapolations. *Management Science*, **38**, pp. 1394–414.

Collopy, F. & Armstrong, J. (2000) *Another error measure for selection of the best forecasting method: the unbiased absolute percentage error*. Working paper: http://www.marketing.wharton.upenn.edu/forecast/paperpdf/armstrong-unbiasedAPE.pdf

Fildes, R. & Makridakis, S. (1995) The impact of empirical studies on time series analysis and forecasting. *International Statistics Review*, **63**, 289–308.

Fildes, R., Hibon, M., Makrdiakis, S. & Mead, N. (1998) The accuracy of extrapolative forecasting methods: additional empirical evidence. *International Journal of Forecasting*, **14**, 339–58.

Foundation for Performance Measurement (1999) Abridged from website: www.fpm.com

Kahraman, G., Ruan, D. & Tolga, E. (2002) Capital budgeting techniques using discounted fuzzy versus probabilistic cash flows. *Information Sciences Journal*, **142**, 1–4.

Kirkham, R.J., Boussabaine, A.H., Grew, R.J. & Sinclair, S.P. (1999) Forecasting the running costs of sport and leisure centres. *Durability of Building Materials and Components 8: Service Life and Asset Management*, Vol. 3, *Service Life Prediction and Sustainable Materials*. National Research Council. Canada Press, Ottawa, Canada, pp. 1728–38.

Makridakis, S. & Hibon, M. (1979) Accuracy of forecasting: an empirical investigation (with discussion). *Journal of the Royal Statistical Society*, Series A, **142**, 97–145.

Makridakis, S., Andersen, A., Carbone, R., *et al.* (1982) The accuracy of extrapolation (time series) methods: results of a forecasting competition. *Journal of Forecasting*, **1**, 111–53.

Makridakis, S., Weelwright, S.C. & McGee, V.E. (1993) *Forecasting: Methods and Applications*, 2nd ed. John Wiley and Sons, New York.

Massheder, K. & Finch, E. (1998) Benchmarking methodologies applied to UK facilities management. *Facilities*, **16** (3/4), 99–106.

Mead, N. (2000) Evidence for the selection of forecasting methods. *Journal of Forecasting*, **19**, 515–35.

Montgomery, D.C. & Runger, G.C. (1994) *Applied Statistics and Probability for Engineers*. John Wiley and Sons, New York.

Myers, C.L. (1971) Forecasting electricity sales. *The Statistician*, **20**(3), 15–22.

Plenty, T., Chen, W. & McGeorge, W. (1999), Accrued value assessment – a dynamic approach for investment appraisal and facilities management. In *Proceedings of 8th International Conference on Durability of Building Materials and Components*, 30 May–3 June, Vancouver, Canada, pp. 1765–73.

Pouliquen, L.Y. (1970) Risk analysis in project appraisal. International Bank for Reconstruction and Development. *World Bank staff occasional papers 11*. Johns Hopkins University Press, Baltimore.

Van Horne, J.C. (1995) *Fundamentals of Financial Management*. Prentice-Hall, Harlow.

Wanous, M. (2000) *A neurofuzzy expert system for competitive tendering in civil engineering*. Unpublished PhD thesis, University of Liverpool.

WBDG (1993) *Whole Building Design Guide*. http://www.wbdg.org/DesignPrinciples.asp?Principles=3&Introduction=3

Yokum, J.T. & Armstrong, J.S. (1995) Beyond accuracy: comparison of criteria used to select forecasting methods. *International Journal of Forecasting*, **11**, 591–7.

5 Whole Life Risk Analysis Techniques

5.1 Introduction

In Chapter 2 we discussed the synergy of risk management and whole life-cycle costing and this chapter aims to consolidate the themes by considering in more detail the structure, purpose and techniques of whole life risk analysis. Dealing with risk can be a complex and time-consuming process, with a wide range of techniques and procedures available to carry out the task. It is the intention here to explain why an appreciation of risk techniques throughout the whole life of a building must be achieved, and what processes and procedures are available by which to do this. This chapter places an emphasis on the qualitative and quantitative techniques available to the analyst during a risk analysis, and where the application of these techniques is most suited. This chapter also covers in detail the concepts of probability theory and the representation of risk through probability distributions. It is hoped that by the end of this chapter, the reader will have a clear idea of the importance of risk analysis techniques to WLCC.

5.2 Risk analysis

There is no single widely accepted definition of risk analysis. It is not a discipline or a methodology, but rather a systematic aid to decision making. The essence of risk analysis is that it attempts to capture all feasible options that have been identified in the risk identification exercise and to then analyse the various outcomes of any decision. Risk analysis provides the ability to gain an insight into what happens if any investment decision, for example, does not proceed according to plan. However, the experience and intuition of the analyst are just as important as the techniques used, as interpretation and ultimate sanction of the information elicited is an important issue.

The development of effectual risk management strategies will identify the most effective environment for risk analysis within the overall modelling system. Simply analysing risk and then managing it is not sufficient (Ward & Chapman 1991). Risk analysis estimating and modelling imposes a discipline to risk identification and evaluation that the traditional systems lack. Essentially, what is being proposed is that when identifying the data that is required, it is essential that a clear picture of the final goals and objectives of the risk management procedure is defined (Tah & Carr 2001).

Formal methods of risk analysis need to be used to ensure that some kind of consistency and uniformity (Kahneman *et al.* 1986) is achieved but many of these techniques are often complicated, time consuming and expensive and consequently the use of these for many projects is prohibitive. Lack of knowledge and doubts as to suitability within the built environment disciplines have also been acknowledged (Akintoye & MacLeod 1997) as reasons for the slow take up. The techniques of risk analysis for use in WLCC might be divided into three categories:

- Deterministic techniques (based on numerical computation of risk)
- Qualitative (techniques that use subjective scoring techniques)
- Quantitative (statistical and probabilistic approaches to quantification).

This division of these concepts is illustrated in Table 5.1. The table shows the most popular techniques of risk analysis that are currently in use, by division. In this chapter we will review the qualitative and quantitative techniques identified in the table that are most appropriate for application within the WLCC process. We will then look at some of the deterministic approaches towards the end of the chapter.

Table 5.1 Techniques for treating uncertainty and risk in the economic evaluation of building investments (Marshall 1999).

Deterministic	Qualitative	Quantitative
Conservative benefit and cost estimating	Risk matrix	Input estimates using probability distribution
Break-even analysis	Risk registers coefficient of variation	Mean–variance criterion
Risk-adjusted discount rate	Event trees (qualitative)	Decision tree analysis
Certainty equivalent technique	SWOT analysis	Simulation (Monte Carlo/ Latin hypercube simulation)
Sensitivity analysis	Risk scoring	Mathematical/analytical technique
Variance and standard deviation	Brainstorming sessions	Artificial intelligence
Net present value	Likelihood/consequence assessment	Fuzzy sets theory
		Event trees (quantitative)

5.3 Qualitative risk analysis

Qualitative risk analysis is where the likelihood or the magnitude of the consequences of an event or occurrence is expressed in qualitative terms (i.e. not quantified by mathematical procedures) as opposed to quantitative risk analysis where the probability or frequency of the outcomes can be estimated and the magnitude of consequences is quantified. In practice, qualitative analysis is often used first to obtain a general indication of the level of risk associated with a project. The results of this analysis may indicate that subsequently it may be necessary to undertake a more specific and detailed quantitative analysis where risk has been identified as particularly important.

Qualitative analysis uses non-numeric or descriptive scales to describe the degree of possible outcomes and the probability of occurrence. These scales can be adapted or adjusted to suit particular circumstances, and numerous descriptions may be used for different risks. Qualitative analysis is used:

(1) As an initial screening activity to identify risks which require more detailed analysis
(2) Where the level of risk does not justify the time and effort required for a more complete analysis, *or*
(3) Where the numerical data is inadequate for a quantitative analysis.

There is a diverse range of techniques available to the analyst when attempting to establish a risk-based approach to WLCC. There are no set rules regarding the selection of the most appropriate technique, as most analysts will tailor their approach to the particular needs and characteristics of the organisation or situation under study. The intention here is to explain the techniques available and where they are best suited in the WLCC analysis.

5.4 Risk matrices

The risk matrix is possibly one of the most basic and widespread techniques in simple risk analysis. It is however generally seen more as a tool for risk identification and risk scoring only, and not for a complete risk analysis, as opposed to some of the more comprehensive techniques that will be discussed later in this chapter. Some may argue that the risk matrix is in fact at the core of the risk management documentation for a project. Certainly, the construction of a risk matrix is a fundamental part of the PFI procurement process (OGC 2003) and is required in all projects.

The composition of a risk matrix usually comprises the following procedures:

(1) Identification of the risks involved in the project
(2) An evaluation of the influence of these risks on the project
(3) An assessment of the likelihood of such risks arising
(4) The calculation of the financial impact (and ranges of possible outcomes).

The format of a risk matrix is usually that of a simple table which is used to assign a score to each identified risk; this can then be used to assist with the risk management process in general. For each of the identified risks, a score is assigned for the probability and impact facets of each risk. Very often a range of 1 to 5 is used for each aspect (1 representing low risk through to 5 representing high risk). Different organisations use different ratings, and as mentioned earlier, the analyst will probably want to tailor the approach to suit the specific circumstances of the project. A simple example of a risk matrix is given in Fig. 5.1.

In this example, the x-axis represents the likelihood of an event occurring, with 1 being very unlikely through to 5 being very likely. The y-axis is a representation of the perceived magnitude of the expected outcome, where 1 is an inconsequential outcome through to 5 which indicates a severe outcome. The matrix is then used to calculate what are referred to as 'risk numbers'; these are obtained by finding the product of these representations of likelihood and severity. An alternative technique to this is shown in Fig. 5.2 where

Consequence of risk (magnitude)	Likelihood of risk				
	1	**2**	**3**	**4**	**5**
1	1	2	3	4	5
2	2	4	6	8	10
3	3	6	9	12	15
4	4	8	12	16	20
5	5	10	15	20	25

Fig. 5.1 Example risk matrix.

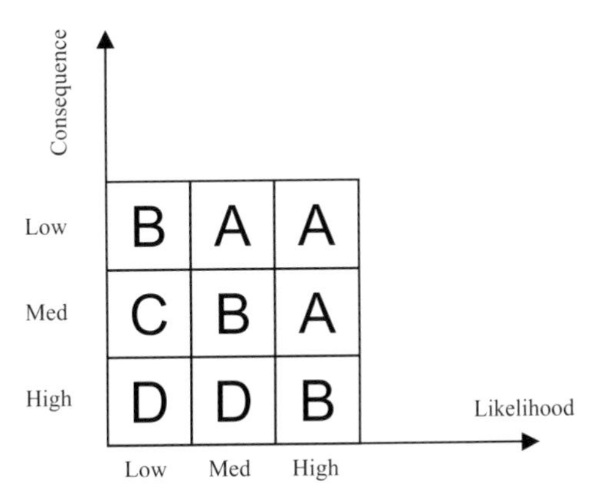

Fig. 5.2 Risk identification scoring.

the 'prioritisation' of risks is defined as a series of situation definitions that are scored according to the risk priority:

A – Indicates intolerable risk which must be removed or reduced immediately
B – Indicates risk which is not altogether intolerable but which should be removed or reduced at the earliest opportunity. Normally, target dates should be set for implementing the countermeasures so that unacceptable delays do not occur
C – Indicates risk which is tolerable but which should be removed or reduced when time allows
D – Indicates risk which may be accepted as it is.

5.5 Risk registers

Unlike risk matrices, which are produced during the early stages of the project, the risk register is a dynamic document. The risk register (sometimes referred to as a risk log) is established during the initial phases of the project and subsequently maintained throughout the whole life-cycle in order to scrutinise potential risks to the success of the project. It should be considered an important component of the organisation's risk management framework. It covers issues or events which may have an adverse effect on that success. These are likely to fall into the following broad areas (Public Records Office 2003):

(1) Legal/regulatory (compliance with legislation and internal procedures or policies)
(2) Quality (e.g. assurance procedures – are they fit for purpose? Are quality auditors knowledgeable and reliable?)
(3) Financial (e.g. contingency costs, agreed budget, budget control)
(4) Contractors (e.g. do they know the specifications? proven record?)
(5) Communication, knowledge and information (involves all stakeholders: consistency? same information communicated to all?)
(6) Political (external and internal: may include interdepartmental relationships and relationships with other bodies)
(7) Commercial (e.g. revenue generation?)
(8) Environmental (e.g. impact on local community).

Like the risk matrices, the risk register assigns each risk with a rating. However, the rating of each risk is not based upon the techniques used in the matrices, but instead involves many factors such as:

(1) Probability of manifestation of the risk
(2) The perceived impact of the risk on the project
(3) The time in which mitigating action must be taken in the event of its occurrence
(4) A statement outlining the case why some risks may be defined as unmanageable
(5) The criticality of the risk in terms of success of project.

Risk	Likelihood L	Likelihood M	Likelihood H	Consequence L	Consequence M	Consequence H	Score
Project scope	✓					✓	
Project definition	✓	✓				✓	
Project objectives		✓			✓		
Dependence of project on other systems			✓			✓	

Fig. 5.3 Example risk register.

Each of these factors may be classed on a scale, perhaps resembling the examples given in the risk matrices section, where 1 is lowest and 5 is highest. The product when these are multiplied together is the rating for each risk. High ratings will then be more important than low ratings. Risks may be displayed in rating order in the register for impact. A tabulated format is helpful. Figure 5.3 shows an example risk register.

The importance of ensuring that the register is constantly updated cannot be overstated. The register should be kept up to date with new risks (potential or actual), which may be added as they emerge, or the rating of each may be adjusted as the project progresses.

The register may also list risk reduction methods (to be used to prevent a risk from occurring), contingency plans (to be invoked if a risk does occur), any costs, which may be involved in mitigation of the risk, and owners for each risk (these need to exist to ensure that risks are avoided or reduced). Should a risk occur, its progress should also be tracked in the register.

When the risks have been identified, it could also be advisable to prepare initial responses to each identified risk, particularly if the risks that are identified require very urgent attention. The analysis may be concluded however during this phase if the assessment immediately suggests a way in which many identified risks can be mitigated. It may be necessary to revisit the identification phase after the assessment phase to see if any consequential 'secondary' risks can be identified: a secondary risk may result from a proposed response to an initial risk and might therefore lead to the response being unsuccessful. The necessity of doing this will largely depend on the size and/or complexity of the project.

5.6 Event trees

Event tree analysis (ETA) is based on binary logic, in which an event either has or has not happened. The analysis involves the visual representation of all the events, which can occur in a project, and these resemble the branches of a tree. It is a valuable technique in analysing the consequences arising from a risk or undesired event. Event trees can be both qualitative and quantitative in nature, the former involving descriptive representation of risk, the latter dealing generally with the numerical probability of outcome.

An event tree must always begin with an initiating event, for example a decision to commit investment funds into new capital. The consequences of the event are followed through a series of possible paths. Each path is assigned with a description of the outcome of the risk should it be realised (qualitative), or the probability of occurrence and the probability of the various possible outcomes can be calculated (quantitative). The goal of a quantitative event tree is to determine the probability of an event based on the outcomes of each event in the chronological sequence of events leading up to it. In considering all possible consequences, the percentage of outcomes which lead to the desired results can be determined.

5.7 Influence diagrams

Influence diagrams are a relatively recent innovation in decision analysis and have proven to be highly useful for risk analysis, risk management and risk communication. An influence diagram is a simple visual representation of a decision problem and is quite similar to a flow chart, with arrows and nodes characterising various stages and procedures within a project (see Fig. 5.4). Influence diagrams can provide an intuitive method of identifying and recording the essential elements, including decisions, uncertainties and objectives, and how these influence each other within a project scenario. Sometimes influence diagrams are called 'knowledge maps'. Like event trees, influence diagrams can be quantitative as well as qualitative, and in the more formal examples influence diagrams show the structure of conditional probability among many related variables that contribute to a given uncertainty; decisions can be analytically solved like decision trees.

Influence diagrams are also used to develop cognitive maps of how experts view a given risk management decision, and how various stakeholders might view the same decision. The differences among expert and lay views form the basis for defining the content of risk communication messages. This view of influence diagrams has been termed the 'mental models' approach by

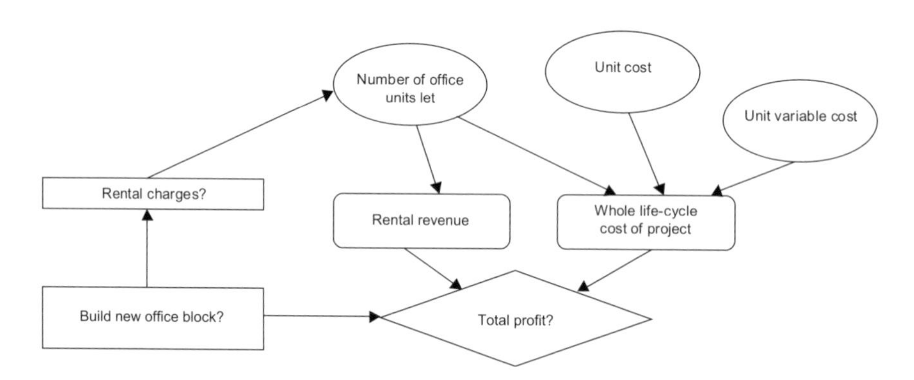

Fig. 5.4 Influence diagram.

a group of researchers at Carnegie Mellon University. Still other uses of influence diagrams include structuring the assumptions, variables and scenarios used when eliciting probabilities from technical experts. Judgement-based probabilities comprise still another aspect of decision analysis that is particularly important for risk management decisions.

The contributing factor diagram (CFD) (Koller 1999) is a useful tool in risk identification as it enables the user to continually update the scenario as new risks become apparent. The CFD is a hybrid of the more formally constructed influence diagram, which uses a mix of symbols to represent decision nodes of various types. Strictly speaking, the CFD is used as a format for assessing the dependencies between certain variables within a risk model, and the discovery of the factors that directly or indirectly affect each other in the project as a whole.

5.8 SWOT analysis

Although not referred to in the literature as a classic risk analysis tool, the SWOT (strengths, weaknesses, opportunities and threats) procedure is a very simple and effective means of carrying out a basic analysis of the risks in a project. Allied to risk matrices, the SWOT analysis could possibly be the first step in establishing a business plan for a specific capital investment decision. It could also be used at any other time during the whole life of a project to quickly identify the risks and benefits of a particular decision. Some key pointers to include in the SWOT analysis could include:

- *Strengths* What are the strengths of the project? What makes it better than other projects? Are these strengths being sufficiently exploited? Are they being sufficiently defended?
- *Weaknesses* What are the weaknesses of the project? What makes it inferior to other ideas? Are there strategies that should be adopted to offset these weaknesses? Should these weaknesses be removed completely?
- *Opportunities* What external factors exist that could be harnessed if appropriate resources are obtained and utilised?
- *Threats* What external factors could threaten the success and viability of the project?

5.9 Brainstorming sessions

Brainstorming is a term used to describe either a group session or an individual attempt to gather as many ideas and possible solutions to a problem as possible. The results to some extent depend upon on the purpose of the brainstorming session and the type of individuals participating in the session.

The key to success lies in the ability of the facilitator (the session leader) to elicit the ideas and viewpoints of the session participants from various angles and to explore each one in greater detail. A successful brainstorming session will end with a solid action plan that can be implemented, analysed and finally assessed. Brainstorming is a useful tool for decision making, especially when attempting to identify the risks that are inherent in a particular project.

The facilitator should discuss the reason for the session and clearly define the objectives and problems associated with the project. Ensuring that all participants agree to the objectives set out and understand the required outcome is vital. Brainstorming sessions are generally a more formalised version of the standard project meetings that are characteristic of almost any project in the built environment.

5.10 Quantitative risk analysis

The applications of quantified risk modelling techniques, which implicitly represent statistical and probabilistic theory, are not widespread in practice, although well-developed techniques have been in existence for some time. As mentioned previously, qualitative techniques are more widely used, with a stronger emphasis on risk identification as opposed to quantification. The complexity of the techniques available may account for this, but quantitative methods can yield far more information and facilitate more accurate decision making. Quantitative methods of risk analysis can be grouped into statistical and probabilistic approaches. Statistical approaches use descriptive statistics such as standard deviation and variance to quantify risk, as well as incorporating these with economic performance measures such as NPV. Probabilistic approaches utilise probability distribution functions and simulation techniques to quantify the risk. Table 5.1 lists currently available techniques for quantitative risk analysis. A review of these techniques is presented in the next sections.

5.11 Probabilistic approaches to risk

Representing risk through probability theory has become accepted, particularly with the advent of many software packages and add-ins that facilitate the simple representation of risk. Several techniques have been reported on in the literature for the quantitative analysis of risk through probability theory. Lorance and Wendling (1999) introduce the importance of the accurate construction of risk probabilities in Monte Carlo simulation modelling. However, the intricacies of modelling the assumptions, that combine to form the Monte Carlo model, are not addressed sufficiently. The issue of modelling the assumptions is discussed in more detail in Maio *et al.* (2000), which looks specifically at the construction of probability distributions for construction operations simulation, and considers the applicability of the goodness-of-fit tests. The importance of the correct statistical procedures for modelling distributions in Monte Carlo models is highlighted in Maddalena *et al.* (2001)

where a discussion of the sensitivity of distributions to the output model is presented. The conclusion here is that the distributions should be constructed initially by identifying the most influential set of variables within the model. This discussion leads to wider consideration of the sensitivity analysis of the inputs constructed. Existing techniques fail to identify the minimum set of stochastic inputs to a model required to adequately characterise the output variance.

5.11.1 Modelling the input probability density functions

Most approaches for economic risk analysis use subjective probabilities to describe the uncertainty of input variables when historical data may not be available (Ranasinghe & Russell 1993; Perry & Hayes 1985; Bjornsson 1977). However, when historical cost data is available, the collation of this data and the subsequent modelling of real-world scenarios can give rise to several problems when trying to create valid probability distributions. A simple heuristic technique for assessing the validity of a distribution is to plot a histogram of the data and visually inspect the variance, kurtosis and skewness of the data over the range (Law 1998). This can give a basic suggestion as to which distribution (or family of distributions) best represents the data, but there are several factors which must also be addressed before selecting a possible distribution.

If probability distributions are to be used to create stochastic assumptions of WLCC inputs, then the way data is analysed and transformed into PDFs is of significant interest and importance. Weiler (1965) concluded that many errors in the outputs of simulation models could be traced back to assigning incorrect values to the parameters of a distribution, and indeed the selection of an appropriate distribution.

Where it is possible to collect data on whole life costs on FM cost centres of interest, such data can be used to specify a distribution based on one of the following approaches: a trace driven simulation, an empirical distribution, or a theoretical distribution function (Maio *et al.* 2000). If data is used to define an empirical distribution, the data is grouped to form a frequency histogram, and the resulting information is transferred to the simulation model. However, if the data set is used to fit a theoretical distribution using heuristics and goodness-of-fit techniques, it smoothes the irregularities that prevail and allows the possibility of sampling the extreme values of the distribution. This technique is regarded generally as the best method for performing simulations.

5.11.2 Defining the distribution

Class intervals, or 'bins', are the ranges by which the data is grouped on a histogram. The number and width of each class interval can have a significant bearing upon which distributions best fit the data being represented (when using the chi-square goodness-of-fit test). Some researchers (Montgomery & Runger 1994) have suggested that the number of class intervals should fall somewhere between five and twenty. They suggested that the square root rule should be used to calculate the number of observations. Simply, taking the square root of the number of observations in the data set derives the number of class intervals.

Sturges's rule is reported on as another method of class interval selection (Maio *et al.* 2000). Sturges's rule states that, for n observations, X_i is to be summarised in a frequency distribution, then the number of class intervals for the distribution should be calculated by:

$$K = [1 + \log_2 n]$$

where K = number of bins and n = number of observations, and

$$M = \frac{[X_{max} - X_{min}]}{K}$$

where M = width of class intervals, X_{max} and X_{min} = maximum and minimum values of observations in the data set.

For the data presented in this paper, the selection of class interval rule is not critical as both methods yield similar results for distribution fitting and chi-square evaluation. Figure 5.2 shows the convergence of both rules between 35 and 55 observations, which reflects the number of observations used in this research. However, for modelling using in excess of 70 observations, careful consideration is needed of which rule is most appropriate as the graph shows a clear divergence of class intervals after that point.

5.11.3 Types of continuous probability density functions

The process of selecting the probability distribution can sometimes present difficulty to the modeller. There is no hard and fast method of assigning a correct distribution. Fortunately, software does now exist to enable the correct choice of distribution from a given data set but a manual method of distribution selection is proposed in Flanagan and Norman (1983):

(1) List everything that is known about the variable and the conditions surrounding the variable
(2) Understand the basic types of probability distribution[s] [and functions]
(3) Select the distribution that [best] characterises the variable under consideration.

This procedure, although quite concise in concept, appears to underestimate the skill and intuition required in modelling accurate distributions. A wide range of techniques and methods are required to construct probability distributions correctly. This chapter will move to address in more detail the statistical methods of selecting the best fitting distribution. However, this next section, by means of a brief narrative, will review the most common types of continuous probability distributions in cost modelling and service life modelling. It should be noted here the argument discussed in Law (1998), which expressed concern for the implications of specifying a theoretical distribution when clearly the data does not provide for an adequate fit. Approaches for dealing with this situation are discussed in Swain *et al.* (1988), which introduces the four-parameter distributions.

The normal distribution (alias: Gaussian)

$$f(x) = \frac{1}{\sqrt{2\pi s}} e^{-(\varsigma - m)^2/2s^2} \quad \text{for } -\infty < \varsigma < \infty$$

Parameters: mean *(m)*, standard deviation *(x)*
The normal distribution is considered the single-most important distribution in probability theory. This distribution (which is a parent distribution of several other continuous types including lognormal) can represent many continuous variables and is particularly useful for representing the uncertainty in inflation and discount rates.

The beta distribution (alias: power function)

$$f(x) = \begin{cases} \dfrac{\left(\frac{x}{s}\right)^{(\alpha-1)} \left(1 - \frac{x}{s}\right)^{(\beta-1)}}{\beta(\alpha, \beta)} & \textit{if}: 0 < x < s; \alpha > 0 \\ & \beta > 0 \\ 0 & \textit{otherwise} \end{cases}$$

where: $\beta(\alpha, \beta) = \dfrac{\Gamma(\alpha)\Gamma(\beta)}{\Gamma(\alpha + \beta)}$
and where: $\Gamma(\beta)$ is the gamma function

Parameters: alpha, beta and scale *(x)*
The beta distribution is very flexible in the nature of forms it can assume but this is reflected to a significant extent in the complexity of the calculation of the parameters. It is commonly used to represent variability over a fixed range. It was identified for use in construction cost simulation in Wall (1997). One of the more important uses though of the beta distribution is as a conjugate distribution for the parameters of the Bernoulli distribution. In this kind of application, the beta distribution is used to represent the uncertainty in the probability of occurrence of an event. It is also used to describe empirical data and predict the random behaviour of fractions and percentages.

The Pareto distribution

$$f(x) = \begin{cases} \dfrac{\beta \cdot L^{\beta}}{x^{(\beta+1)}} & if\,(x > L) \\ 0 & if\,(x \leq L) \end{cases}$$

Parameters: location (L), shape (β)
The Pareto distribution finds itself mainly involved in the representation of empirical phenomena in finance such as personal incomes, stock price fluctuations, etc. The potential of the distribution is considered in several civil engineering applications including in the statistical analysis of extreme wind speeds upon buildings (Holmes & Moriarty 1999) and for econometric procedures in Creamer (1997).

The Weibull distribution (alias: Rayleigh distribution)

$$f(x) = \begin{cases} \left(\dfrac{\beta}{\alpha}\right)\left(\dfrac{x-L}{\alpha}\right)^{\beta-1} e^{-\left(\frac{x-l}{\alpha}\right)^{\beta}} & if\,(x \geq L) \\ 0 & if\,(x < L) \end{cases}$$

Parameters: location (L), scale (α) and shape (β)
The Weibull distribution is perhaps essential to research as this deals with life expectancy, life-cycles, future forecasting and deterioration of elements. For instance, in Kirkham *et al.* (2002a) the Weibull distribution was recognised as the most suitable distribution for simulation of electricity cost forecasts in hospital buildings. This distribution is considered part of a family of distributions that assume the properties of several other distributions. For example, it can be used to model the exponential and Rayleigh distributions among others. The parameters above dictate the form that the distribution takes. For instance, when the shape parameter is equal to 1.0, then the Weibull distribution is equal to the exponential distribution.

The gamma distribution (aliases: erlang, exponential and chi-square)

$$f(x) = \begin{cases} \dfrac{\left(\dfrac{x-L}{\beta}\right)^{\alpha-1} e^{-\left(\frac{x-L}{\beta}\right)}}{\Gamma(\alpha)\beta} & if\,(x > L) \\ 0 & if\,(x \leq L) \end{cases}$$

Where $\Gamma(\alpha)$ is the gamma function

Parameters: location (L), shape (α) and scale (β)
The gamma distribution can represent a wide range of quantities and is related to several other distributions such as lognormal, exponential, Pascal, Erlang, Possion and chi-square. The gamma distribution is also used to measure the

time between the occurrence of events when the event process is not completely random (Decisioneering 2000). It is also used quite comprehensively in economic and financial risk theory. In Milevsky (1998) the probability density function of the present value of a stochastic rate of interest is considered, and proof that its inverse is gamma distributed is presented. The results found are considered useful for computing the discount rate distribution, in a real world stochastic environment, under a fixed probabilistic confidence level.

The exponential distribution

$$f(x) = \begin{matrix} \lambda e^{-\lambda x} \; if\, x \geq 0 \\ 0 \;\; if\, x < 0 \end{matrix}$$

Parameters: rate (λ)
The exponential distribution is related to the discreet distribution, Poisson. Its use mainly lies in the modelling of random points in time of events (Kulkarni 1995). For example, the distribution could be used to model the probabilities that the structure of a building may fail at a point in time. The exponential distribution however exhibits a special characteristic which some describe as 'memoryless'. Basically, this distribution exhibits the property that time has no effect on future outcomes. Markov models, for instance, exploit the properties of the exponential distribution; this is characterised in Boukas and Yang (1999) where a study of the exponential stability of Markov models is discussed.

The lognormal distribution

$$f(x) = \frac{1}{x\sqrt{2\pi s}} e^{-(\log_e(x) - m)^2/2s^2} \text{ for } 0 < x < \infty$$

Parameters: log mean (m), log standard deviation (s)
The lognormal distribution is generally used to represent data sets that are positively skewed, that is say the distribution is centred around the lower end values in the range. There are three underlying conditions to the lognormal distribution:

(1) The uncertain variable can increase without limits but cannot fall below zero
(2) The uncertain variable is positively skewed
(3) The natural logarithm of the uncertain variable yields a normal distribution.

The general rule of thumb for using the lognormal distribution is that if the coefficient of variability is greater than 30%, then the lognormal distribution is valid, otherwise the normal distribution should be used. The lognormal distribution was reported on in Kirkham *et al.* (2002b) where it was identified as the best distribution for modelling facilities management costs in WLCC models for NHS hospital buildings.

5.11.4 Validating the selected distribution

Whichever method of calculating the number of class intervals in the distribution is used, the next stage is to fit a distribution to the data set. Although visual inspection can reveal which kind of distributions are most likely to represent the data, a statistical test should be performed to validate the choice of selected distribution.

The *chi-square test* is a formal comparison of the relationship between the observed data set and the theoretical distribution fitted. The chi-square test is however highly correlated to the class interval rule chosen (either the square root rule or Sturges's rule) and as such, the method used to calculate the number of class intervals affects the chi-square test results, particularly so in data sets where more than 70 observations are used. This has led to the conclusion that the chi-square test is weakened by its dependence on the class interval rule.

Notwithstanding, this test is widely used by construction researchers involved in fitting distributions to data sets.

The chi-square statistic is defined as:

$$\chi^2 = \sum_{i=1}^{K} \frac{(N_i - E_i)^2}{E_i}$$

Where:
K = number of bins
N_i = the number of observed samples in the ith bin
E_i = the expected number of samples in the ith bin.

The *Kolmogorov-Smirnov (K-S) test* compares the observed data with a hypothesised distribution. The strength of the K-S test is that it is indifferent to the class interval rule. The test works by assessing the likelihood of the observed data originating from a hypothesised distribution with the estimated parameters. It has been noted that the K-S test should be used in preference to the chi-square test as its result is not affected by the class interval rule; it is valid for the exact sample size and has more power against alternative distributions.

The K-S test is defined as:

$$D_n = \sup[\left| F_n(x) - \hat{F}(x) \right|]$$

Where:
n = total number of observations
$\hat{F}(x)$ = the fitted cumulative distribution function
$F_n(x) = \frac{N_x}{n}$
N_x = the number of X_is less than x.

The *Anderson-Darling (A-D) test*, although similar to the K-S test, lends more weight to the tails of the distribution. The A-D test was designed to detect discrepancies in the tails of the fitted distribution and observed data.

The A-D test is defined as:

$$A_n^2 = n\int_{-\infty}^{+\infty}[F_n(x) - \hat{F}(x)]^2\psi(x)\,\hat{f}(x)\,dx$$

Where:
$\hat{f}(x)$ = the hypothesised density function
$\hat{F}(x)$ = the hypothesised cumulative density function
$F_n(x) = \dfrac{N_x}{n}$
N_x = the number of X_is less than x.

5.12 Simulation

The expression 'Monte Carlo method' is very general. Monte Carlo (MC) methods are stochastic techniques – meaning they are based on the use of random numbers and probability statistics to investigate problems. MC methods are used in a variety of quantitative situations from economics to nuclear physics to regulating the flow of traffic movements. The way MC techniques are applied varies widely from field to field, and there are dozens of subsets of MC methods, even within engineering situations for example. Strictly speaking, to express a situation or modelling procedure as a 'Monte Carlo' experiment, random numbers must be used to examine the problem.

Monte Carlo simulation is a statistical technique by which a quantity is calculated repeatedly, using randomly selected 'what if' scenarios for each calculation. Although the simulation process is internally complex, commercial computer software performs the calculations as a single operation, presenting results in simple graphs and tables. These results approximate the full range of possible outcomes, and the likelihood of each. When Monte Carlo simulation is applied to risk assessment, risk appears as a frequency distribution graph similar to the familiar bell-shaped curve, which non-statisticians can understand intuitively.

To describe mathematically a MC analysis, assume a general mathematical model with an outcome, Y, that is a function of uncertain inputs, X_i, expressed as (Maddalena *et al.* 2001):

$$y = f(X_1, X_2, \ldots, X_m)$$

Because of the nature of Monte Carlo sampling, the outcome distribution of Y, $F(Y)$, is subject to a certain degree of imprecision. The amount of imprecision or scatter in the $F(Y)$ depends mainly upon the number of outcomes, m, used to generate the distribution given as:

$$n = p(1 - p)\left(\frac{z_{(1-\alpha)}}{\Delta_p}\right)^2$$

Where p is the percentile of interest, $Z_{(1-\alpha)}$ is the standard normal deviate for the $(1-\alpha)$ probability and Δ_p is the range encompassing p with a $(1-\alpha)$ degree of certainty (Morgan & Henrion 1998). The imprecision in the empirical cumulative distributions from repeated MC analyses is shown in the example in Fig 5.1. However, the use of Monte Carlo simulation requires the analyst to be aware of the important limitations associated with its implementation as a risk assessment tool. Firstly, most available software cannot distinguish between variability and uncertainty. Some factors, such as energy cost and building occupancy, show well-described differences among data sets. These differences are called 'variability'. Other factors, such as frequency and duration of maintenance and repair regimes, are simply unknown. This lack of knowledge is called 'uncertainty'. Current Monte Carlo software treats uncertainty as if it were variability, which may produce misleading results.

Ignoring correlations among exposure variables can bias Monte Carlo calculations. The importance of the assessment of correlations in Monte Carlo methods is outlined in Wall (1997) who draws attention to the possibility of the ignorance of correlations leading to serious misassessment of risk. However, information on possible correlations is seldom available. Exposure factors developed from short-term studies with large populations may not accurately represent long-term conditions in small populations.

Secondly, the tails of Monte Carlo risk distributions, which are of significant interest, are highly sensitive to the shape of the input distributions. Because of these limitations, this has led to some researchers recommending that Monte Carlo simulation should not be the sole, or even primary, risk assessment method. Nevertheless, Monte Carlo simulation is clearly superior to the qualitative procedures currently used to analyse uncertainty and variability.

5.12.1 Latin hypercube simulation

The Latin hypercube sampling procedure in Monte Carlo simulation is an alternative method of generating assumption values during a simulation (Decisioneering 2000). This method works by segmenting the assumption's probability distribution into a number of non-overlapping intervals, each having equal probability (for goodness-of-fit methods, see the equal-probability chi-square test).

Then, from each interval, the simulation procedure selects a value at random according to the probability distribution within the interval. The collection of these values forms the Latin hypercube sample. The analyst is required to specify the sample size option to control the number of intervals.

Latin hypercube sampling is generally more precise for producing random samples than conventional Monte Carlo sampling, because the full range of the distribution is sampled more evenly and consistently. Latin hypercube sampling (Iman *et al.* 1974) has been shown to require fewer model iterations to approximate the desired variable distribution than the simple Monte Carlo method. The Latin hypercube technique ensures that the entire range of each variable is sampled. A statistical summary of the model results will produce indices of sensitivity and uncertainty that relate the effects of heterogeneity of input variables to model predictions. In addition, frequency distributions of model state variables can be compiled to determine the relative frequencies of the most frequent and maximum classes resulting from the model simulations.

5.12.2 Calculating the number of iterations in Monte Carlo simulation: the stopping rule

One of the issues in Monte Carlo simulation that has not been sufficiently addressed is the number of iterations required in a simulation. Evidence from the literature suggests that researchers have tended to make a subjective assessment of the number of iterations required in a simulation, but it is clear that this will affect the results of the simulation. Schuyler (1997) presents one method of defining the number of iterations required in a simulation by way of the standard error of the mean statistic, given by:

$$\xi = \frac{\alpha}{\sqrt{n}}$$

Where:
ξ = standard error of mean
α = standard deviation of the variable of interest
n = number of iterations in the trial.

The author suggests that running the simulation until ξ is less than 1% of the mean is a good rule-of-thumb stopping rule. The author also reports on the improvements in ξ reduction by using a similar number of Latin hypercube simulation (LHS) iterations; 100 LHS iterations achieved the same value of ξ as 10 000 Monte Carlo iterations. Wainwright (1999) offers an improvement to the rule offered by Schuyler, using 'bootstrapping' methods to identify the required number of iterations.

(1) Determine a 'sample size' (for Latin hypercube sampling and for correlation sampling). Perhaps this should be the lesser of: (a) 100, or (b) $\frac{1}{5}$ (rounded) of the minimum number of trials anticipated.
(2) At 'sample size' intervals, determine whether adequate convergence has been achieved. For most purposes we want adequate confidence in the mean value, such as +/−1% with 68% confidence. This is determined by using the 'bootstrapping' technique:

- Randomly select, say, 'sample size' values (sampling these without replacement would be best)
- Determine the sample mean of these values
- Repeat steps (1) and (2), say, 50 times
- Compute the standard deviation of the sample means from step (2); this approximates the standard error of the mean (SEM)
- Stop when the SEM is <1% of the mean.

5.13 Sensitivity analysis

Sensitivity analysis measures the impact on project outcomes of changing one or more key input values (assumptions) about which there is uncertainty. In engineering economics, Marshall (1999) discusses sensitivity analysis as the measurement of economic impact resulting from alternative values of uncertain variables that affect the economics of operating and maintaining a building. Sensitivity analysis is normally performed altering one assumption at a time, but it is possible to adjust more than one assumption simultaneously.

Insufficient data for adequately estimating environmental model parameters introduce uncertainty into model predictions. In order for complex models to provide reliable predictions of environmental effects for long time periods over broad spatial scales, a technique is needed to provide insight into the inadequacies of parameter estimation. Monte Carlo sampling methods have been used to randomly vary parameters over a range of values from a specified frequency distribution, and generate corresponding sets of model predictions. Sensitivity/uncertainty analyses of these parameter sets and model predictions have been used to relate parameter variance to model predictions.

5.14 Markov theory

The preceding sections of this chapter have dealt with applied probability theory for risk analysis. This section discusses the application of another form of applied probability theory, which can be used to incorporate risk analysis in service life forecasting and planning. This special type of stochastic procedure, which has been reported on in the literature, is known as Markovian chain theory. The Markovian process can be very powerful in probabilistic service life modelling; however the mathematics can be equally difficult. In this section the aim is just to give a flavour of the technique and how it can be applied with a WLCC model.

5.14.1 The Markov chain

Markov chains are a special kind of stochastic process in that they exhibit what is known as the memoryless property. The memoryless property states that the future condition of a building element is conditional only on its current state and is independent of the past condition (or state) of the element. This is a known as a conditional probability and can be expressed in its most basic form by:

$$P(A) = \{A \mid B\}$$

Where $P(A)$ is the probability of the system and $\{A \mid B\}$ is the statement which says that the probability of A is conditionally dependent upon B. Therefore, relating this basic principle to the Markov property, a Markov chain can be expressed as (Wirahadikusumah *et al.* 1999):

$$P(X_{t+1} = i_{t+1} \mid X_t = i_t, X_{t-1} = i_{t-1}, \ldots, X_1 = i_1, X_0 = i_0)$$
$$= P(X_{t+1} = i_{t+1} \mid X_t = i_t)$$

5.14.2 Preconditions of the Markov process

In most circumstances and in order to reduce the complexity of the modelling process, it is assumed that the future condition of a building element is dependent only upon the present state, and independent of the past condition (the Markov property – see section 5.14.1). It can therefore be said that for all states i and j and all t,

$$P(X_{t+1} = i_{t+1} \mid X_t = i_t)$$

is independent of t. The probability, p_{ij}, that the building condition is in state i at time t and it will be in state j at time $t + 1$ does not change, that is to say it remains stationary over time, unless of course rehabilitation is performed or there are other external factors which are not included in the assessment change. The stationary assumption is that:

$$P(X_{t+1} = i_{t+1} \mid X_t = i_t) = p_{ij}$$

5.14.3 Transition probabilities

Not only does the Markov property facilitate a risk integrated approach to forecasting service lives in buildings, it also allows the analyst to define the point where the condition of a component will move from one predefined state to another. So, for example, you may wish to know when a component moves from the state 'good' to the state 'poor'. Transition probabilities in Markov chain theories are normally computed from the n step transition probability matrix. From the matrix, the Chapman-Kolmogorov equations are used to solve the matrix. This is demonstrated in Kulkarni (1995) where the n step transition probabilities satisfy the following equation:

$$p_{ij}^{(n)} = \sum_{r \in S} p_{ir}^{(k)} \, p_{rj}^{(n-k)} \quad (i, j \in S)$$

Where k is a fixed integer such that $0 \leq k \leq n$.

5.15 Deterministic measures of risk

The final group of risk analysis techniques are the deterministic approaches. These techniques, which do not involve any consideration of probability theory, have been widely used for decades in investment appraisal. These techniques principally utilise the variance and standard deviation functions to quantify risk.

5.15.1 Variance and standard deviation as a measure of risk

The variance, or the standard deviation, measures the dispersion of forecasted values about the mean or expected value. In WLCC, this gives rise to the simple problem, what is to be expected? In forecasting, it is not known what will occur in the future, although estimates can be made based upon historical trends. The variance provides information on the extent of the possible deviations of the actual return from the expected return. The variance therefore from a distribution is given by the formula:

$$\sigma_x^2 = \sum_{i=1}^{n} P_i(x_i - Ex)^2 = E(x - Ex)^2$$

Where:
$E(x)$ = the forecasted value
x_i = the ith possible outcome
P_i = the probability of obtaining the ith outcome x_i
n = the number of possible outcomes.

In using this to compare investment decisions, for example options A and B, if A is a more risky investment than B, then in terms of the variance $\sigma_A^2 > \sigma_B^2$ its standard deviation must also be greater. For the purposes of ranking mutually exclusive projects, the variance or standard deviation may be used interchangeably.

5.15.2 Variance and standard deviation of NPV as a measure of risk

The last section gave an example of how variance can be used as an elementary measure of risk. However, in whole life costing the element of time is important to the forecast and so this calculation must be compared with a time period of discounted costs. It follows then that the variance of the NPV can give us an indication of the risk distribution for a given period of time (in this case, the service life).

When dealing with variance as a measure of risk over a service life period, it is important though to consider some of the following issues.

- How should the variance within the service life be handled? Should the measure of risk reflect the sum or average of the yearly variances?
- How can risk at different intervals within the service life be treated?
- How should the variance between the years be handled?

It is known that in the case of major capital investment, after the initial outlay, for many years a negative cash flow will exist, dependent upon the magnitude of the investment. However, as an example, consider the investment in a new energy saving heating unit for a ward. Assuming that after the initial capital outlay the following two years' cash flows will be positive as a result of net savings on energy, then:

$$E(NPV) = \alpha EX_1 + \alpha^2 EX_2 - I$$

Where:
EX_1 = the expected value of net cash flow in year 1
EX_2 = the expected value of net cash flow in year 2
I = initial capital investment
$\alpha = 1/(1 + r)$ = the coefficient for capitalising cash flow over the time period, where r denotes the 'risk less' interest rate:
$E(NPV)$ = expected NPV

Given the assumption that the net cash flow for the second year is statistically independent of the first, the variance of the NPV can then be defined as:

$$\sigma^2 = \sum_i \sum_j P_i P_j [\alpha X_{1i} + \alpha^2 X_{2j} - I - (\alpha EX_1 + \alpha^2 EX_2 - I)]^2$$

Where:
X_{1i} = denotes net cash flow in year 1, with probability Pi
X_{2j} = denotes net cash flow in year 2, with probability Pj.

This equation shows that the variance of the net present value depends on all possible combinations of cash flows in years 1 and 2, multiplied by their probability of occurrence, which equals the product P_iP_j in the case of independence. Hence, the variance measure now reflects the nature of whole life costing in that the period of time is accounted for.

5.16 Mathematical and analytical techniques

Estimation and allocation of WLCC can be viewed as an optimisation problem. It might be possible to formulate this problem as:

> What is the optimal allocation of whole life costs among a building asset's whole life budget centres for a predetermined level of risk?

If WLCC and risk are stated in this fashion, then the concept of whole life costing and risk analysis might be modelled as an optimisation and minimisation of cost and risk respectively. Precisely this notion of cost optimality and risk minimisation might be formulised by mathematical models such as utility theory and other operational research techniques. The purpose of such formulisation is to model WLCC decisions criteria in order to minimise risk and maximise NPV among WLCC decision alternatives with equal risk. One may argue that there is a utility function that assigns a numerical value to each WLCC alternative. As most WLCC decisions are expressed in monetary terms, the utility function may have profit or other benefits as argument, measuring the satisfaction obtained from investments.

For example, suppose that the WLCC of a building asset of component C is measured by V(c) and its riskiness by R(c). Then, risk-cost utility models imply the following form of the utility function U(.), where the component with the lowest WLC of U(.) will be chosen from the set of alternatives.

$$U(c) = f(V(c), R(c))$$

The above function f(c.) quantifies the trade-off between WLCC and risk. In this respect WLCC is measured by the expected monetary value and the riskiness by the variance.

As stated in Chapters 2 and 3, risk and uncertainty are important issues in the context of achieving whole life objectives. For this reason, mathematical optimisation models might be used to evaluate whole life alternatives and associated risks: This might include:

- Maximisation of NPV
- Maximisation of asset financial performance
- Minimisation of whole life risks
- Maximisation of life service of asset and components.

Among other analytical tools for risk analysis that may have an impact on WLCC risk assessment is the analytical hierachy process (AHP) method. AHP is a mathematical method for analysing complex decision problems with multiple criteria, originally developed by Saaty (2001). AHP is measurement theory based on mathematical and psychological foundations. AHP can deal with qualitative risk attributes as well as quantitative ones. It has been widely used in the quantification and ranking of risk attributes. It also has the potential to be combined with the Delphi technique when assessing the risk outcomes of whole life aspects of building assets. AHP can be used in the overall priorities of WLCC decision alternatives. For more details on the AHP theory, readers are referred to Saaty (2001).

5.17 Artificial intelligence and fuzzy set theory

A risk-based approach to WLCC analysis requires that the analyst uses advanced technologies that can effectively and efficiently identify factors that prevent project stakeholders from achieving their objectives. As explained in Chapters 2 and 3, WLCC risk analysis is a very complex problem caused by difficulties in defining the cause–effect relationship between the risk determinant factors. These difficulties lie in the complex challenge that the WLCC analyst faces in collecting data from several different sources into a coherent WLCC risk assessment model. Among the advanced techniques that show promise as enablers of WLCC risk analysis are ANN, fuzzy set theory, case-based reasoning, genetic algorithms and a combination of these techniques.

5.17.1 Artificial neural networks (ANNs)

ANNs are specialised applications of artificial intelligence technology. Derived from models of how the brain is organised and how it learns new information, these networks are trained to make decisions. ANNs have many advantages over traditional methods of modelling in situations where the process to be modelled is complex to the extent that it cannot be explicitly represented in mathematical terms or that explicit formation causes loss of sensitivity due to oversimplification. ANN systems can provide precise, non-linear correlation between their input and output data, but the mechanism underlying that correlation is opaque. Although expert systems are programmed with detailed algorithms for making choices (a reasonable method when the relationships between variables can be clearly specified), ANNs are merely provided with the values of both independent and dependent variables and allowed to determine the relationships between predictors and outcomes. They are immune to many of the errors in judgement that are characteristic of humans. Thus their decisions are not adversely affected by failure to consider base rates or covariance. Self-interest, conflicts and bias play no role in their functioning.

ANNs are classifiers by nature, offer the capacity to simultaneously consider multiple types of evidence and can assist the whole life cost analyst in assessing risks and making whole life judgements. ANNs are proved to be superior to other modelling approaches in cases where data is available in sufficiently large samples, the range of values to be analysed for each case is large, and the underlying associations among the data are fuzzy and ill defined. ANNs provide several advantages over traditional advanced statistical techniques such as logistic regression. It is proven that unlike traditional models for risk analysis, ANNs are non-linear and do not require any prior assumptions about the distribution properties of the underlying whole life data. ANNs learn from the cases analysed by constructing an input-output mapping. ANNs learn the patterns that are evident in WLCC risks and create a knowledge base for prediction or classification of WLCC risk factors. Risk knowledge is presented in ANNs by their structure and activation function or state (weights), and the network can readily interpolate and in some cases extrapolate from the existing state to solve unseen problems.

However, ANN parameters (i.e. weights, learning rules, transfer functions, topology, etc.) reveal nothing than can rationally be interpreted as a causal explanation of the real world relationship modelled by the trained network. This opacity problem has two effects on ANN technology. First, it reduces confidence in ANN technology. Secondly, it makes the design of ANN systems ad hoc based. Another problem with ANN systems is that with nominal or ordinal representation of input and output, useful information could be disregarded. However, combining ANN systems with qualitative causal models can solve this problem of opacity. The most common approach to combining qualitative causal models with ANN systems is the neurofuzzy approach. Combining neural network systems with fuzzy models helps to explain their behaviour and to validate their performance.

ANN application in WLCC risk analysis can be divided into two categories. First, it may be possible to develop ANN models that can classify and rank WLCC risk factors at any stage of the life-cycle of building assets. Second, ANN can be used as a risk-forecasting tool at the operation stage. Here, operating data is usually available annually. This data might be used to develop an ANN time series model to forecast operation risks or costs. Other areas where ANN may be of use in WLCC is risk systems modelling.

5.17.2 Fuzzy set theory

Realistically whole life cost estimators are more interested in the direction of movement of costs rather than its exact forecast value, and they would be interested not only in novel efforts in forecasting but also in proactive forecasting strategies. Fuzzy logic application to decision making might help in this process. Fuzzy models are particularly suited to making decisions involving new technologies where uncertainties inherent in the situation are complex.

The magnitude and timing of future service life of building components and associated WLCC need to be carefully estimated, taking into consideration the related risks and uncertainties. The major problem that whole life analysts encounter in making financial decisions involves both the uncertainty and the ambiguity surrounding expected WLCC results. In the case of complex projects the problems of uncertainty and ambiguity assume even greater proportions because of the difficulty in predicting the impact of unexpected changes on building assets and consequently on WLCC computation. The uncertainty and ambiguity are caused not only by project-related problems but also by economic and technological factors, as explained in Chapters 2 and 3. Whole life costs are generally spread over the life span of assets and are dominated by building assets-related constraints and by environmental (external) constraints over which whole life managers have little control. As has been shown above, several methods have been used in risk analysis. None of these approaches address the problem of ambiguity in the whole life process because events and even probabilities associated with them are not always mutually exclusive. The answer to this problem might be found in fuzzy theory. The fuzzy set approach has been widely applied to help in real life decision making processes.

Fuzzy logic is finding wide popularity in various applications that include management, economics and engineering. The theory was introduced by Zadeh (1965) over three decades ago, but only recently has its application received large momentum. Fuzzy logic deals with uncertain or imprecise situations and decisions. Although fuzzy theory deals with imprecise information, it is based on sound quantitative mathematical theory. A variable in fuzzy logic has sets of values, which are characterised by linguistic expression, such as very high risk, average risk, low risk, etc. These linguistic expressions are represented numerically by fuzzy sets. Every fuzzy set is characterised by a membership function, which varies from 0 to 1. Fuzzy sets have a distinct feature of allowing partial membership. A given element can be a member of a fuzzy set, with degree of membership varying from 0 (non-member) to 1 (full member), in contrast to crisp or conventional sets, where an element can either be or not be part of the set. Linguistic variables as described by Zadeh (1965) provide a means of modelling human tolerance for imprecision by encoding decision relevant information into labels of fuzzy sets. Fuzzy technologies are an approximation that can be used to model WLCC decision processes for which mathematical precision is impossible or impractical.

Fuzzy set theory provides a convenient way of representing the following whole life concepts:

- When WLCC probabilistic data for risk assessment is extremely rare and insufficient, the utilisation of subjective judgement data based on expert's experiences and fuzzy concepts
- Use of fuzzy membership curves to model uncertainties range in risk factors and whole life estimates
- Use of fuzzy set theory in assessing constructional and operational risks
- Use of fuzzy set theory in assessing and estimating service life risks
- Use of fuzzy linguistic variables to describe imprecise whole life risk factors
- Subjective fuzzy probability can be used to represent WLCC risks and estimates
- Fuzzy sets can be used to represent whole life costs when historical data is not available to define the underlying statistical distribution.

5.18 Summary

This chapter has presented a brief introduction to the concepts of risk in cost forecasting and modelling. Risk is a confusing term in construction and built environment research. For some researchers, it refers to the act of identifying and dealing with risk such as health and safety and management failure. In the construction industry, most professionals are exposed to the qualitative techniques we discussed such as risk registers and matrices. However, risk is also a mathematical science that encompasses the mathematical modelling of uncertainty, particularly in cost modelling procedures. This is vital in the application of forecasting techniques as it allows the analyst to determine the likelihood of projected values within predetermined ranges.

This chapter has also presented a review of applied probability theories in quantification of risk in cost forecasting and component service life prediction. From this review so far, it is clear that a WLCC model not only requires the ability to forecast service lives accurately to determine the analysis period, but it should also incorporate a stochastic process to deal with the uncertainty inherent in such a process. The Monte Carlo methods discussed can also be a significant part of the WLCC model. The determination of overall WLCC and the risks associated with the forecast are derived from a Monte Carlo simulation. This chapter has aimed to clarify the issues surrounding the use of simulation and the importance of correctly defining the distributions associated with assumptions in WLCC models.

References

Akintoye, A.S. & Macleod, M.J. (1997) Risk analysis and management in construction. *International Journal of Project Management*, **15**(1), 31–8.

Bjornsson, H.C. (1977) *Risk Analysis of Construction Cost Estimates*. Transactions of the American Association of Cost Engineers, pp. 182–9, West Virginia, USA.

Boukas, E.K. & Yang, H. (1999) Exponential stabilisation of stochastic systems with Markovian jumping parameters. *Automatica*, **35**(8), 1437–41.

Creamer, E. (1997) A characterisation of the generalised Pareto distribution with an application to re-insurance. *Insurance: Mathematics and Economics*, **21**(3), 249.

Decisioneering (2000) *CB Predictor User Manual*. Decisioneering Inc, Denver, Colorado.

Flanagan, R. & Norman, G. (1983) *Life-Cycle Costing for Construction*. Surveyors Publications, Royal Institution of Chartered Surveyors, London.

Holmes, J.D. & Moriarty, W.W. (1999) Application of the generalised Pareto distribution to extreme value analysis in wind engineering. *Journal of Wind Engineering and Industrial Dynamics*, **83**(1–3), 1–10.

Iman *et al.* (1974) quoted from *Latin Hypercube Sensitivity Analysis* by M. Lynn Tharp. http://www.cped.ornl.gov/cad_cp/text/lts.html

Kahneman, D., Slovic, P. & Tversky, A. (1986) *Judgment Under Uncertainty: Heuristics and Biases*. Cambridge University Press, Cambridge.

Kirkham, R.J, Boussabaine, A.H. & Kirkham, M.P. (2002a) Stochastic time series forecasting of electricity costs in an NHS acute care hospital building, for use in whole life-cycle costing. *Engineering, Construction and Architectural Management Journal*, **9**(1), 38–52.

Kirkham, R.J., Boussabaine, A.H. & Awwad, A.H. (2002b) Probability distributions of facilities management costs in NHS acute care hospital buildings. *Construction Management and Economics Journal*, **20**, 251–61.

Koller, G. (1999) *Risk Assessment and Decision Making in Business and Industry*. CRC Press, New York.

Kulkarni, V.G. (1995) *Modelling and Analysis of Stochastic Systems*. Chapman and Hall, London.

Law, A.M. (1998) *Expert Fit User Manual*. Averill M. Law and Associates, Arizona.

Lorance, R.B. & Wendling, R.V. (1999) Basic techniques for analysing and the presentation of cost risk analysis. *Proceedings of 1999 AACE International Annual Conference*, Washington.

Maddalena, R.L., McKone, T.E., Hsieh, D.P.H. & Geng, S. (2001) Influential input classification in probabilistic multimedia models. *Stochastic Environmental Research and Risk Assessment (SERRA)*, **15**(1), 18.

Maio, C., Schexnayder, C., Knutson, K. & Weber, S. (2000) Probability distribution functions for construction simulation. *Journal of Construction Engineering and Management*, **126**(4), 285–92.

Marshall, H.E. (1999) *Technology Management Handbook*, Chapter 8.12. CRC Press, New York.

Milevsky, M.A. (1998) The present value of a stochastic perpetuity and the Gamma distribution. *Insurance: Mathematics and Economics*, **20**(3), 243–50.

Montgomery, D.C. & Runger, G.C. (1994) *Applied Statistics and Probability for Engineers*. John Wiley and Sons, New York.

Morgan, M.G. & Henrion, M. (1998) *Uncertainty: A Guide to Dealing with Uncertainty in Quantitative Risk and Policy Analysis*. Cambridge University Press, New York.

OGC (2003) *Private Finance Initiative Material*. Office for Government Commerce, www.ogc.gov.uk

Perry, J.G. & Hayes, R.W. (1985) Risk and its management in construction projects. *Proceedings of the Institution of Civil Engineers*, **78**(1), 499–521.

Public Records Office (2003) *What is a Risk Register?* Public Records Office, http://www.pro.gov.uk/archives/A2A/riskregister.htm

Ranasinghe, M. & Russell, A.D. (1993) Elicitation of subjective probabilities for economic risk analysis: an investigation. *Construction Management and Economics*, **11**, 326–40.

Saaty, T.L. (2001) *The Analytic Hierarchy Process for Decisions in a Complex World*. RWS Publications, Pittsburgh.

Schuyler, J. (1997) *Monte Carlo Stopping Rule, Part 1*. http://maxvalue.com/tip025.htm

Swain, J.J., Venkatraman, S. & Wilson, J.R. (1988) Least squares estimation of distribution functions in Johnson translation systems. *Journal of Statistical Computer Simulation*, **29**, 271–97.

Tah, J.H.M. & Carr, V. (2001) Towards a framework for project risk knowledge management in the construction supply chain. *Advances in Engineering Software*, **32**(10–11), 835.

Wainwright, E. (1999) *Monte Carlo Stopping Rule, Part 2*. http://maxvalue.com/tip029.htm

Wall, D.M. (1997) Distributions and correlations in Monte Carlo simulation. *Construction Management and Economics*, **15**, 241–58.

Ward, S.C. & Chapman, C.B. (1991) Extending the use of risk analysis in project management. *International Journal of Project Management*, **9**(2), 117–23.

Weiler, H. (1965) The use of the incomplete beta functions for prior distributions in binomial sampling. *Technometrics*, **7**(3), 335–47.

Wirahadikusumah, R., Abraham, D.M. & Castello, J. (1999) Markov decision process for sewer rehabilitation. *Engineering, Construction and Architectural Management*, **6**(4), 358–70.

Zadeh, L.A. (1965) Fuzzy sets. *Information and Control*, **8**, 338–53.

Part II
Whole Life-cycle Costing: The Design Stage

6 Design Service Life Planning

6.1 Introduction

The service life of new and existing structures represents an essential parameter, which must be considered when there are decisions to be made regarding future investment (Gosav 1999). Service life of buildings is an important factor in whole life-cycle cost analysis and the assessment of global costs (Flourentzou *et al.* 1999).

The process of service life planning can be applied to existing buildings as well as new constructions. For existing buildings and components, many of the choices have been predetermined and the building is already some way through its service life. Therefore planning the service life will focus on assessing the residual service life of the components and optimising planned maintenance and costs of replacements. For new assets, the application of some form of mathematical assessment of the service life of components is required. These assessments are made upon an a priori assumption of the lifetime of the system, for example from estimated service lives provided by the manufacturer of components.

In estimating the service life of components, decision makers should consider the effect of several influencing factors. It is not only the question of how long a building component can last, but also how long will a built environment asset be retained? All built environment assets and their components have widely different service lives, depending on several factors such as the physical, economic, functional, technological, social and legal obsolescence. Service life of building assets is influenced by decisions taken through the entire life-cycle of assets, from planning to maintenance to disposal. That is why the accuracy of service life cannot be exactly determined unless all influencing factors are taken into account at the design stage. It is also important that all stakeholders should understand all the design decision consequences.

Several mechanisms are currently in place that can assist professionals in arriving at component and building service life prediction (Hovde 2002). This chapter presents a review of both stochastic methods and deterministic methods of service life prediction; it then introduces an innovative method for determining the remaining life of building components.

To ensure that key definitions are understood, the following definitions have been extracted from ISO 15686-1 and Nireki & Motohashi (2002):

Service life: period of time after installation during which a building or its parts meets or exceeds the performance requirements.
Service life planning: preparation of the brief and design for the building and its parts to achieve the desired life. Period of time during which all essential performance characteristics of a properly maintained product, component or building in service exceed the minimum acceptable values.
Design life: service life intended by the designer.

6.2 Estimation of service life for new structures

Generally, methods of predicting the service life of building components can be classified in two distinct categories:

- Deterministic prediction methods
- Stochastic prediction methods.

An accurate simulation of whole life costs must be based on robust prediction methods of the service life of new building assets or the remaining service life of existing buildings and their components. Martin *et al.* (1994) presented a set of criteria for selecting an appropriate service life prediction method. These attributes include:

- Ability to deal with a large variability in times to failure for components
- Analysis of multivariate data
- Ability to discriminate among service life determinant variables
- Ability to provide mathematical techniques to predict service life
- Links with existing WLCC and LCA tools.

6.2.1 Deterministic prediction methods for service life estimation

Twenty years ago, predicting the service lives of building materials and components was only a distant vision (Frohnsdorff 1996). Today, the possibility of incorporating predictions of services lives into the design process and whole life-cycle costing methods is becoming evident. Since the early 1990s knowledge about service life of building materials, components and buildings has been developed in many countries and different types of service life prediction methods have been suggested. This work is mainly done under the auspice of the International Organisation for Standardisation (ISO). ISO has published a standard, ISO 15686 Part 1 (ISO 2000), that deals with the service life of building and constructed asset facilities. A considerable section of the ISO standard is devoted to the factor method for determining the service lives of components. The standard focuses on service life planning where the objective is to assure, as far as possible, that the estimated service life of the building or components will be at least as long as its design life. As the length of the service life cannot be known precisely in advance, the objective becomes to make an appropriately reliable forecast of the service life using available data. The output of the process is a series of predicted service lives of components, and a projection of maintenance and replacement needs and timings.

One of the main features of the ISO standard is the description of the service life prediction method based on the factor method. The method is based on similar factorial methods that have been developed by the Architectural Institute of Japan. The factor method is developed to estimate the service life of building components or assembly in specific conditions, assuming these conditions remain static through the whole life-cycle of facilities. The method is based on reference service life and modifying factors that relate to the specific conditions of the component under analysis. The factor method is a deterministic model expressed in the following relationship:

$$ESLC = RSLC \times A \times B \times C \times D \times E \times F$$

Where:
$ESLC$ = estimated service life
$RSLC$ = reference service life
A = material/component factor
B = design factor
C = workmanship factor
D = in use factor
E = external environment factor
F = maintenance factor.

The major problem highlighted in the use of this method is the fact that no recognisable national or international standard existed for the production of accurate RSLCs (Bourke & Davis 1999). Other drawbacks are that the service life is treated as a deterministic value. But, in the real world, service life has large scatter and should be modelled as a stochastic quantity (Siemes & Edvardsen 1999). The estimation of attributes in the above expression is based on subjective assessment of perceived condition of service of components. The resulting forecast should also be treated as a distribution rather than a point forecast. It is feasible to develop probability density functions for building components using supplier data, testing or estimates from previous experience. If data and experience are unavailable, a probability density curve of the estimated service life can be defined by professional estimate or by the recursive Delphi method. Because of the way elements and components are integrated into building design, plus their interaction with each other in a dynamic environment that is constantly changing, service life prediction is always going to be uncertain no matter what type of data and information is collected about the process and product. Hence, it is better if some effort is also invested in the development of tools and strategies that help to mitigate the risk associated with the process.

6.2.2 *Stochastic processes in service life forecasting*

Stochastic processes are processes that move over time in a probabilistic manner (Wirahadikusumah *et al.* 1999). A stochastic process is defined as an indexed collection of random variables $\{X_t\}$, where the index t moves through a given set of T. Often, T is taken to represent a set of non-negative integers, and X_t represents a measurable characteristic.

Stochastic models have been widely reported on in literature with particular application to deterioration models and failure prediction applications in engineering systems. Madarat *et al.* (1995), in their work on estimation of transition probabilities (the process of an element moving from one condition state to another), highlighted the problems associated with the ad hoc approaches to estimating these probabilities. An econometric method was proposed for estimating transition probabilities based upon condition rating data. The method is based on probability technique, which recognises the relationship between the latent nature of the components' performance and explicit links with deterioration of the components to relevant explanatory variables. This work was further supported by work reported in DeStefano and Grivas (1998) who developed a method for estimating the transition probabilities of condition states in a bridge and applying these to an existing infrastructure deterioration management system. The researchers found that the inclusion of probabilistic methods significantly increased the accuracy of the deterioration model. Markovian models for bridge maintenance management were presented in Scherer and Glagola (1994), who identified the strength of the Markov process in decision making techniques.

Butt *et al.* (1987) presented the development of a pavement prediction model using a Markov process and compared it with the curve-fitting techniques. A probability-based Markov model was developed, in order to represent the probabilistic behaviour and rate of deterioration of pavements. The Markov process imposes a rational structure on the deterioration model.

A probabilistic model for tunnelling projects using Markov processes has been discussed by Touran (1997) who proposes that states of work and non-work for a tunnel-boring machine can be modelled using a Markov chain. A general probabilistic model is proposed for calculating the cumulative distribution function of the total length that can be tunnelled at any one time. The author draws attention specifically to the problems of dealing deterministically with certain parameters, and highlights the potential benefits of a distribution of values as opposed to definite values, so that costs, for example, may be defined by choosing a reasonable confidence level.

Application of Markov processes in maintenance and inspection models is discussed in Williams and Hirani (1997). They describe a model which determines the optimal multilevel inspection-maintenance policy for a stochastically deteriorating multistate subsystem. The subsystem in this model is assumed to be a non-decreasing semi-Markov process. Similarly, Fung and Makis (1997) present a Markov renewal theory based model for assessing deterioration and failure times in machinery components.

Aarseth and Hovde (1999) and Moser (1999) presented methods for treating the factor method attributes as stochastic inputs and thus provided a probability distribution of the service life prediction, with an associated confidence interval determined by the analyst. These distributions of forecasted service life are generally expressed as three parameters, the expected service life plus/minus one standard deviation of the mean.

As a starting point for stochastic estimation of design service life of building assets one can use the design lives of works and construction products advocated by the European Union in its technical approvals publication (EOTA 1999) and the suggested minimum design lives for components from ISO 2000. Figure 6.1 shows graphically some of the service life for building systems advocated by client forums and cost consultants in the UK (Bartlett 2002). Based on the services life recommended by these publications, decision makers can use their experience to develop a range of estimates for service lives. The recommendations from these two organisations and the range estimate are shown in Tables 6.1 and 6.2. The range estimate shown in Table 6.2 is based on the following philosophy. The design service lives stated by ISO and EOTA are used as a base service lives estimate and represent the most likely lives based on the available information. Thus, a range around the base design lives estimate is a measure of variation in the base time.

For example, a range of −5% to +15% will be modelled, as shown in Fig. 6.2. It can be seen that the base service life can be reduced by 5% and increased by 15%, the most likely change being 0%. It is important that decision makers should not estimate ranges that place the most likely service life in the base estimate as the median of the range, as shown in Fig. 6.2. For example, estimators may be tempted to offer +10% as shown in Fig. 6.2. The problem with this type of range estimate is that it tends not to shift the base service life because the most likely occurrence is no change to the base

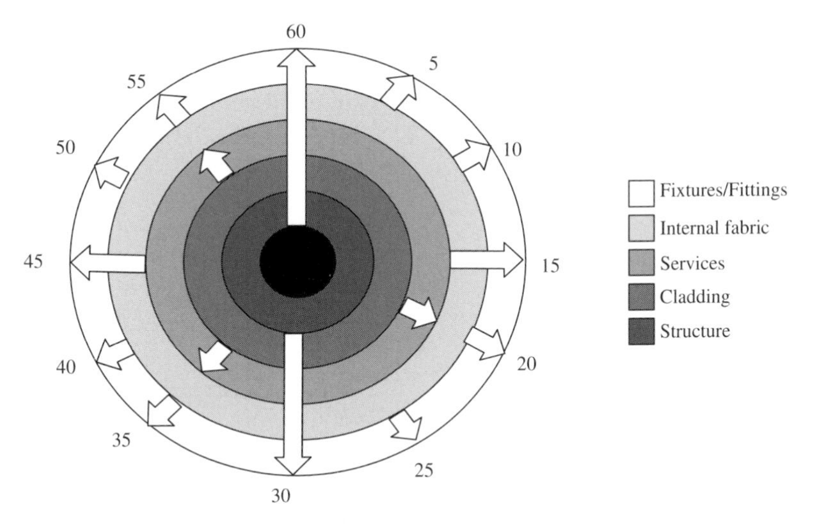

Fig. 6.1 Benchmarking service life planning for building components (Bartlett 2003).

Table 6.1 Design lives of works and building components for different types of buildings (from EOTA (1999) and Hovde (2002)).

Assumed working life of works		Assumed working life of construction products (years)		
Category	Years	Repairable/easily replaceable	Less easily repairable/ replaceable	Lifetime of works
Short	10	10	10	10
Medium	25	10	25	25
Normal	50	10	25	50
Long	100	10	25	100

service life (Noor 2000). Decision makers should consider all possibilities including extreme conditions and should try to define realistic ranges. In the example given in Fig. 6.2 a continuous probability distribution is used for the definition of the range. The range defines the limits of the distribution, and the most likely effect is no change to the base service life estimate. What type of continuous distribution is best suited to represent the range? This is largely dependent on the type of problem to be modelled. There might be situations where decision makers prefer to use discrete probability distribution for modelling range estimates. It should be noted that there are also several mathematical functions that can be used to model the variation in service life estimate.

Information for developing this kind of range estimate can be extracted from several sources, for example, building maintenance information, manufacturer databases, condition surveys and expert knowledge. As the range estimate is based on assumptions, it is important that these assumptions are checked at different key milestones of the whole life decision making process, as described in Chapter 2, to make sure that circumstances have not changed and assumptions are still valid. This type of information can then be fed into

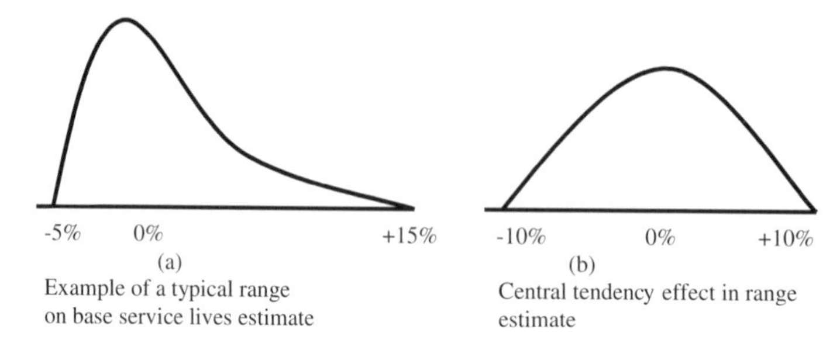

Fig. 6.2 Range estimate for service life planning.

Table 6.2 Suggested minimum design services lives for components and the use of estimate ranging.

Design life of building					Inaccessible or structural components					Components where replacement is expensive or difficult					Major replaceable components					Building services				
Min. variation as %		Most likely	Max. variation as %		Min. variation as %		Most likely	Max. variation as %		Min. variation as %		Most likely	Max. variation as %		Min. variation as %		Most likely	Max. variation as %		Min. variation as %		Most likely	Max. variation as %	
%	Min.		%	Max.	%	Min.		%	Max.	%	Min.		%	Max.	%	Min.		%	Max.	%	Min.		%	Max.
		Unlimited					Unlimited					100					40					25		
90	135	150	120	180	95	142	150	130	195	80	80	100	110	110	75	30	40	120	48	70	17	25	110	27
		100					100					100					40					25		
		60					60					60					40					25		
		25					25					25					25					25		
		15					15					15					15					15		
		10					10					10					10					10		

simulation software for generating probability distribution for design service lives. Once the best-fit probability is determined it should then be used in WLCC computation.

6.2.3 Example of estimating service life for new structures

The basis of the demonstration example is a façade of a school building, which is south facing. Other façades are treated separately due to variation in service conditions. Table 6.3 shows the assumptions for all relevant service life factors. Their assumed contribution to the service life is based on the range estimate described above and on the factors description given in ISO 2000. The range estimate is approximated by continuous probability density functions. Weibull and beta density functions are used arbitrarily to model the variation in the service life factors. These two distributions are selected due to their ability to

Table 6.3 Assumptions for the façade example.

Service life modifying factor	Service condition	Distribution parameters			
		Distribution	Mean	α	β
Quality of component A(x)	General variation of component with low risk of failure	Weibull	1.86	1	2
Work execution level C(x)	Reasonably good with low risk of defects	Beta	0.76	2	3
Design level B(x)	Good with low risk of errors	Beta	0.8	2	3
Outdoor environment D(x)	Sheltered from rain and frost but high temperature fluctuation with medium risk of extreme conditions	Weibull	1.06	1	1.5
In use conditions E(x)	Occasional access for maintenance with low risk	Beta	0.6	2	3
Maintenance level M(x)	Low maintenance requirement with high risk of not maintained on time	Beta	0.6	3	2
Reference service life Y(0)	Based on manufacture durability data	Weibull	50	30	4

Note: For Weibull distribution α and β are location and shape factors.

represent a wide range of problems. The characteristics of the probability density function of the service life modifying factors for the south façade are shown in Fig. 6.3. A Monte Carlo simulation model is used to extract a probability density function of service life. The results from the simulation are used to find the best distribution that represents the service life of the façade. Based on the KS best-fit method, lognormal distribution is found to be the best fit to this particular problem taking into account the existing assumptions. Figure 6.4 shows the characteristics of the generated distribution. The façade service probability distribution function can now be used as an input in whole life-cycle costing analysis.

6.3 Estimation of the remaining service life for existing structures

Maintenance costs contribute significantly to the total cost of a building over its lifetime. Almost 12% of the total running cost of buildings in the UK is related to maintenance. There is a growing awareness that unplanned maintenance and refurbishment costs may amount to half of all money spent on existing buildings. Estimates of the unplanned cost in UK construction output range from £8 billion to £20 billion per annum (CBBP 2001). Long-term planning

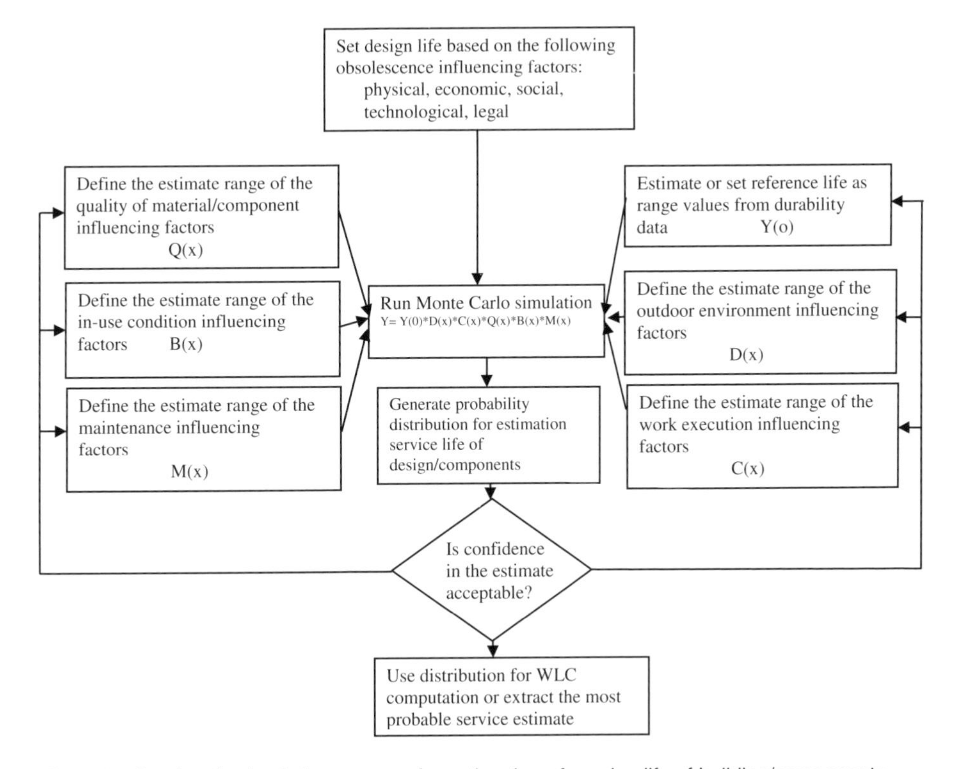

Fig. 6.3 Stochastic simulation process for estimation of service life of building/components.

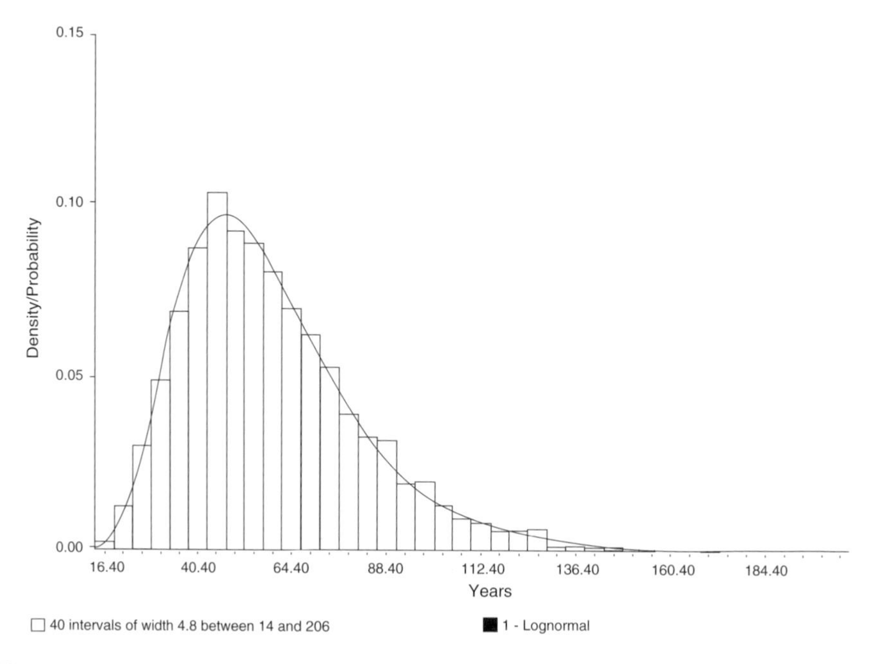

Fig. 6.4 Service life probability distribution of the example.

using WLCC analysis for building maintenance could contribute to the goals of reducing operation and maintenance costs of the building stock, and consequently could contribute directly to the achievement of sustainable built environment. Maintenance models must be based on the life-cycle approach and incorporate the current condition of building components in the prediction of their future condition. By taking account of future condition as a consequence of present rehabilitation action, the best alternative as well as the optimum time for maintenance of components can be determined. By knowing the expected number of components needed to be maintained and rehabilitated, the expected costs as well as their variability can be determined and budgeted for. This will assist maintenance decision makers to obtain a deeper understanding of the life-cycle costs of existing building stock. Also, by knowing the future maintenance costs and their variations, different future possibilities could be sensitively analysed and planned.

Data on the future performance of building components is not always readily accessible; also, this information may be of little use as the key question is concerned with the probable date of repair or replacement, in other words, the point at which the building or component reaches the end of its economic life.

Major building components such as roofing and structural elements require replacement at varying intervals (Bourke & Davis 1999). This mathematically represents a system where the current and future condition of the building or component is totally independent of the past condition. This property is known as a Markovian property. The methodology used here to model the remaining service is based on a special class of Markov systems known as Poisson processes, as described in Severance (2001). Traditionally synchronous problems are modelled by discrete probabilistic Markov systems and solved using the Markovian chain techniques. However, the state of a building component's condition is considered here as a synchronous event driven in continuous time. Therefore, if the transition of states of a building component occurs, this will increment the state of the system by one, but the exact time and probability of this change of state is unknown. If the interevent times are considered to be exponentially distributed, then the process is said to exhibit a special class of the Markovian property, the Poisson property. The following subsections describe the Markovian process.

6.3.1 Methodology

The methodology adopted in the development of the Markov model is illustrated in Fig. 6.5. The following steps were used in the modelling process.

(1) Identification of the building and engineering components. The components were then separated into building and engineering subsections.
(2) Following this, data was elicited from a survey to obtain the estimated service lives and condition ratings of each element.
(3) Elicitation of weights for each building component. This was conducted by means of a survey amongst 10 acute care NHS estate mangers to identify the importance ranking of each component. Using a weighting technique, each element was re-assigned a service life estimate. The condition ratings were kept the same.
(4) Once the weights were calculated, these were then normalised and the estimated remaining life for each component was then recalculated using the normalised weights.
(5) The resulting service life predictions were then summed together for each of the subsections using a weighted average calculation.
(6) The overall remaining life of the building was calculated using the Law of the Minimum of Exponentials.
(7) The resulting remaining life is then deemed to be the period from the current condition state to the final condition state. This is the study period that should be used in WLCC calculation.
(8) Finally, the transition probability was calculated. This quantifies the likelihood of the system moving from the current state into the final state in the calculated time period.

6.3.2 Data elicitation and processing

Many large organisations carry out annual surveys to establish the physical condition of buildings and components within their building stock. For example, in the NHS physical condition survey each building asset is broken down into 19 elements or building components. Each of these elements is assigned a condition rating on a scale of A-D (A = new through to D = imminent failure). For existing elements, condition rating C is acceptable but B is desirable, and

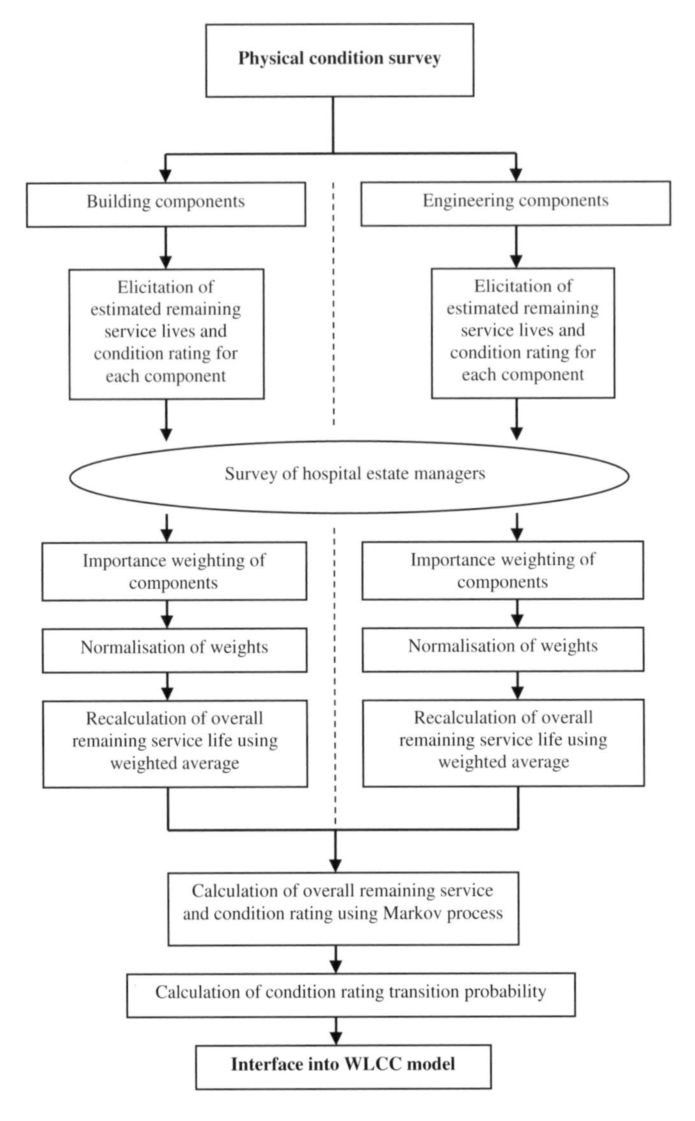

Fig. 6.5 Service life model: development methodology.

so a cost to upgrade to B for any components in C or below is also specified. The analyst must also assign a remaining life estimate; that is, the period of time from the current condition rating to the time when the element enters the condition rating D. A physical condition survey was carried out on the case study building to obtain the initial service life estimates and condition ratings for the building and engineering elements. This was carried out by an estate management expert in accordance with the guidelines laid down in Estatecode (NHS Estates guidance document 1993).

Once the physical condition survey was completed, the next stage of the modelling process was to identify the relative importance of each of the building and engineering components. The validity of existing processes of determining the overall remaining life is questionable due to the assumption that all components should be treated as if of equal importance. Therefore, 10 estate managers from NHS trusts were interviewed and asked to assess on a scale of 1–5 (1 being the most important) each of the 19 building and engineering elements in terms of importance. Importance here is deemed as the level of likelihood of breakdown that the estate managers place upon that element. This is done on a relative scale. From this survey, the final importance ranking scores were elicited. To minimise time and for ease of use, the final overall importance rankings were simply calculated by taking the discrete ranking with the highest frequency for each element. The final overall importance ranking scores are summarised in Table 6.4.

Having calculated the discrete importance rankings for each of the 19 elements, these were recirculated amongst the estate managers to confirm acceptance and a consensus of the scores. The next stage of the modelling process was to assign weights to each of the elements based upon the importance rankings elicited in the estate managers' survey. The rank-sum weights method is used for this purpose. This method is chosen purely in the absence of any published literature on selection of weighting methods in service life planning. The results of the weightings exercise for building and engineering elements are shown in Table 6.5.

Table 6.4 Overall importance ranking of engineering and building elements.

No.	Engineering element	Importance ranking	No.	Building element	Importance ranking
07	Electrical system	1	01	Structure	1
08	Heating system	1	02	External fabric	1
09	Steam system	1	03	Roof	1
10	Ventilation	1	04	Internal fabric	2
11	Phones, paging system	2	05	Int. fixtures and fittings	3
12	Alarms	1	06	External works/ gardens	4
14	Piped medical gases or ventilation and air conditioning	1	13	Drainage	1
15	H or C water	1			
16	Lifts	2			
17	Boiler calorifier	1			
18	Fixed plant or equipment	1			

Table 6.5 Weightings for engineering and building elements.

Engineering element	Rank sum method	Building element	Rank sum method
Electrical system	9.615384615	Structure	18.34862385
Heating system	9.615384615	External fabric	18.34862385
Steam system	9.615384615	Roof	18.34862385
Ventilation	9.615384615	Internal fabric	12.8440367
Phones, paging system	6.730769231	Internal fixtures and fittings	7.339449541
Alarms	9.615384615	External works/gardens	6.422018349
Piped medical gases or ventilation and air conditioning	9.615384615	Drainage	18.34862385
H or C water	9.615384615		
Lifts	6.730769231		
Boiler/calorifier	9.615384615		
Fixed plant/equipment	9.615384615		

6.3.3 *Computation of overall remaining service life*

The overall remaining service life was then recalculated using a weighted average for the engineering and building components. Initially, elements are grouped according to their importance ranking and then the estimated service lives of these components is averaged out and then multiplied by the sum of the normalised weights (in %). This was performed for the building and engineering components and is presented in Table 6.6.

From the previous calculations, it was found that for the main ward block building, the weighted average remaining life for the aggregated engineering elements is 10.16 years [T_1] and for the building elements is 21.40 years [T_2]. The overall condition states are condition rating B for building elements and C for engineering elements. It is known that for the aggregated building and engineering states, the lower of the two condition states determines the overall condition state (i.e. if engineering = C and building = B then overall condition rating = C, and if engineering = B and building = C then overall condition rating = C). If the overall system for calculating the expected remaining life can be expressed as a Markovian process, then this implies that all expected lifetimes must be exponentially distributed. The method of weighted averages used to calculate the expected remaining life for both the aggregated states for building and engineering systems must result in exponentially distributed expected remaining lives. It is also known that the lower of the two condition

Table 6.6 Calculation of the weighted average remaining service life.

	Elements	Condtn state	Import. ranking	Remng life	Raw weight	Normalised weight	Normalised weight (%)
01	Structure	**B**	1	40	100	0.1835	18.348624
02	Ext. fabric	C	1	38	100	0.1835	18.348624
03	Roof	C	1	8	100	0.1835	18.348624
04	Int. fabric	C	2	12	70	0.1284	12.844037
05	Int. fix. or fitt.	**B**	3	12	40	0.0734	7.3394495
06	Ext wks or gdns	**B**	4	16	35	0.0642	6.4220183
07	Elect. system	C	1	9	100	0.0962	9.6153846
08	Htg system	C	1	13	100	0.0962	9.6153846
09	Steam system	C	1	9	100	0.0962	9.6153846
10	Ventilation	C	1	11	100	0.0962	9.6153846
11	Phones, paging system	C	2	8	70	0.0673	6.7307692
12	Alarms	C	1	13	100	0.0962	9.6153846
13	Drainage	**B**	1	12	100	0.1835	18.348624
14	PMG or VAC	C	1	9	100	0.0962	9.6153846
15	H or C water	**B**	1	13	100	0.0962	9.6153846
16	Lifts	C	2	14	70	0.0673	6.7307692
17	Boiler calorifier	**B**	1	8	100	0.0962	9.6153846
18	Fixed plant or equipment	C	1	6	100	0.0962	9.6153846
	Building:	Sum of weights			**545**	**1.000**	**100.000**
	Engineering:	Sum of weights			**1040**	**1.000**	**100.000**
20	**Building**	C		**Straight average**			**19.69**
				Weighted average			**21.40**
21	**Engineering**	C		**Straight average**			**10.21**
				Weighted average			**10.16**

states for both engineering and building elements determines the overall condition state, as described above; it then follows that a stochastic Markovian process known as the minimum of exponentials rule can be used to calculate the estimated overall remaining life of the ward block building.

If this theorem is then related to the expected remaining life calculations for the main ward block building, we have T_1, T_2 as random variables where $E[T_1]$ and $E[T_2]$ are known. $E[T_1]$ = expected remaining service life of the aggregated building components and $E[T_2]$ = expected remaining life of the aggregated engineering components, using the minimum of exponentials theorem. Overall remaining life of the system T is given by:

$$T = min\{T_1, T_2\}, \text{ therefore, E[T] is required}$$

In a Markov process, all expected lifetimes are exponentially distributed, so from the equation for the exponential distribution given as:

$$f(x) = \begin{matrix} \lambda e^{-\lambda x} & if\, x \geq 0 \\ 0 & if\, x < 0 \end{matrix}$$

The engineering and building remaining lives can be expressed as:

$$f_{T_1}(x) = \frac{1}{E[T_1]} \bullet e^{-\frac{x}{E[T_1]}} = \lambda_1 e^{-\lambda_1 x}, \lambda_1 = \frac{1}{E[T_1]}$$

$$f_{T_2}(x) = \frac{1}{E[T_2]} \bullet e^{-\frac{x}{E[T_2]}} = \lambda_2 e^{-\lambda_2 x}, \lambda_2 = \frac{1}{E[T_2]}$$

Where the aggregated building components = f_{T_1}(x), and the aggregated engineering components = f_{T_2}(x). Hence, if $T = \min\{T_1, T_2\}$ is again exponentially distributed then the overall expected remaining life of the system can be expressed as:

$$\lambda = \lambda_1 + \lambda_2 = \frac{1}{E[T_1]} + \frac{1}{E[T_2]}$$

$$\Rightarrow E[T] = \frac{1}{\lambda} = \frac{1}{\frac{1}{E[T_1]} + \frac{1}{E[T_2]}}$$

$$= \frac{1}{\frac{1}{21.40} + \frac{1}{10.16}} = 6.891 \text{ years}$$

Therefore, the overall remaining life of the system is 6.891 years.

In the Poisson process, the calculation of transition probabilities is simplified significantly in that the transition probability can be calculated in a much easier fashion. In continuous time Markov chains, knowing that the exponential distribution of mathematical expectation of remaining life for the system is X years, given in E[T], and that the overall condition state of the system is in state C, then it is possible to calculate the probabilities of the system moving from condition state C to condition state C/D in that calculated period of remaining life. This is given in the following formula, which is given as a function of the mathematical expectation of remaining life:

$$E[T] = \sum_{k=0}^{\infty} k \bullet P\{t = k\} = \sum_{k=0}^{\infty} k \bullet (1 - P_{CD})^k \bullet P_{CD}$$

$$= P_{CD} \bullet (1 - P_{CD}) \bullet \sum_{k=0}^{\infty} k \bullet (1 - P_{CD})^{k-1}$$

$$= P_{CD} \bullet (1 - P_{CD}) \bullet \frac{1}{[1 - (1 - P_{CD})]^2}$$

$$= \frac{1 - P_{CD}}{P_{CD}} = \frac{1}{P_{CD}} - 1$$

Comment:

$$\sum_{k=1}^{\infty} k \bullet \alpha^{k-1} = \frac{1}{(1 - \alpha)^2}$$

$$\alpha \in (0,1)$$

The equation described above can be simply modified to account for the transition probability of the system remaining in the original state as opposed to moving to the final condition state, if this is required.

6.4 Summary

Estimation of service life is an important phase in the process of service life planning. The factor method, modification of reference service life by factors to take into consideration the specific in-use conditions defined in ISO, will remain one of the practical methods at present until another alternative is developed.

This chapter has presented a novel approach to modelling the service life of an acute care hospital building. It uses a combination of weighted averages and a Markov process to determine the overall remaining life of the building and its associated transition probabilities. The results of this interface into the WLCC as the forecasted remaining service life is the analysis period for the model. This is because at the end of the forecasted service life, the system moves from state C to state D, which in terms of the NHS estate is the end of the life of the system. The next chapter looks at the modelling of each of the WLCC cost centres.

References

Aarseth, L.I. & Hovde, P.J. (1999) A stochastic approach to the factor method for estimating service life. In *Proceedings of the 8th International Conference on the Durability of Building Materials and Components.* National Research Council Canada, Vancouver.

Bartlett, E. (2002) Whole life application in cost consultancy, the real cost of building applying whole life costing in construction. Unpublished presentation at one day conference at Royal Institute of British Architects, London.

Bourke, K. & Davis, H. (1999) Estimating service lives using the factor method for use in whole life costing. In *Proceedings of the 8th International Conference on the Durability of Building Materials and Components.* National Research Council Canada, Vancouver.

Butt, A., Shahin, M.Y., Feighan, K.J. & Carpenter, S.H. (1987) Pavement performance prediction model using the Markov process. *Transportation Research Record.* Record 1123, Transportation Research Board, Washington DC.

CBBP (2001) *Factsheet on Whole Life Costing.* www.cbpp.org.uk/resourcecentre/publications/document.jsp?documentID=115776. The Construction Best Practice Programme, Watford.

DeStefano, P.D. & Grivas, D.A. (1998) Method for estimating transition probability in bridge deterioration models. *ASCE Journal of Infrastructure Systems,* **5**(1), 87.

EOTA (1999) *Assessment of Working Life of Products.* European Organisation for Technical Approvals, Brussels.

Flourentzou, F., Brandt, E. & Wetzel, C. (1999) MEDIC – a method for predicting the residual service life and refurbishment investment budgets. *Durability of Building Materials and Components 8: Service Life and Asset Management,* Vol. 3, *Service Life Prediction and Sustainable Materials.* National Research Council Canada Press, Ottawa, pp. 1281–8.

Frohnsdorff, G.J. & Martin, J.W. (1996) *Towards Prediction of Building Service Life: The Standards Imperative, Durability of Building Materials and Components 7,* Vol. 2 (ed. C. Sjostrom). E. and F.N. Spon, London.

Fung, J. & Makis, V. (1997) An inspection model with generally distributed restoration and repair times. *Micro-electron Reliability,* **37**(3), 381–9.

Gosav, I. (1999) Field studies concerning service life prediction. In *Proceedings of the 8th International Conference on the Durability of Building Materials and Components.* National Research Council Canada, Vancouver.

Hovde, P.J. (2002) The factor method for service life prediction from theoretical evaluation to practical implementation. *9th International Conference on Durability of Buildings Materials and Components,* Brisbane, Australia, paper 232.

ISO 15686 (2000) Buildings – service life planning. ISO/TC59/SC14/WG4 Document N29, *Buildings and Constructed Assets: Service Life Planning – Life-Cycle Costing,* International Organisation for Standardisation, Geneva, Switzerland.

Madanat, S.M., Mishalani, R.G. and Wan Ibrahim, W.H. (1995) Estimation of Infrastructure Transition Probabilities from Condition Rating Data. *ASCE Journal of Infrastructure Systems,* Vol. **1**(2), 120–25.

Martin, J., Saunders, S., Floyd, F. & Wineburg, J. (1994) *Methodologies for Predicting the Service Lives of Coating Systems.* NIST Building Science Series 172. National Institute of Standards and Technology. Gaithersburg, Maryland.

Moser, K. (1999) Towards the practical evaluation of service life – illustrative application of the probabilistic approach. *Durability of Building Materials and Components 8: Service Life and Asset Management,* Vol. 2, *Service Life Prediction and Sustainable Materials.* National Research Council Canada Press, Ottawa, Canada.

NHS Estates 1993 (*Estatecode*), NHS Estates, The Stationery Office Books, HMSO, London.

Nireki, T. & Motohashi, K. (2002) Toward practical application of factor method for estimating service life of building. *9th International Conference on Durability of Buildings Materials and Components*, Brisbane, Australia, paper 218.

Noor, I. (2000) Guidelines for successful risk facilitating and analysis. *Cost Engineering*, **24**(4), 32–7.

Scherer, W.T. & Glagola, D.M. (1994) Markovian models for bridge maintenance management. *Journal of Transportation Engineering*, **120**(1), 37–51.

Severance, F.L. (2001) *System Modelling and Simulation: an Introduction*. John Wiley and Sons Ltd, Chichester.

Siemes, T. & Edvardsen, C. (1999) Duracrete service life design for concrete structures: a basis for durability of other building materials and components. In *Proceedings of the 8th International Conference on the Durability of Building Materials and Components*. National Research Council Canada, Vancouver.

Touran, A. (1997) Probabilistic model for tunnelling project using Markov chain. *Journal of Construction Engineering and Management*, **123**(4), 444–9.

Williams, G.B. & Hirani, R.S. (1997) A delay time multi-level on-condition preventative maintenance inspection model based on constant base interval risk – when inspection detects pending failure. *International Journal of Machine Tools Manufacturing*, **37**(6), 823–36.

Wirahadikusumah, R., Abraham, D.M. & Castello, J. (1999) Markov decision process for sewer rehabilitation. *Engineering, Construction and Architectural Management*, **6**(4), 358–70.

7 Design Environmental Life-cycle Assessment

7.1 Introduction

The rapid development of construction industry technology has enabled the transformation of the built environment in many ways, changing the nature and extent of the environmental impacts of construction facilities. Resource depletion, waste, and air and land pollution are some of the examples of the environmental problems which have emerged as a result of exporting waste into the environment. One of the main problems associated with building assets is that they have not only immediate local effect, but may have a more global impact on the environment. This is becoming apparent with the increasing scientific evidence of the cumulative effects of some of the environment impacts over urban space and the health of the population. As a result of these, pressures have been exercised on those responsible for the environmental interventions to improve the environmental performance of the building assets. Among these, the construction manufacturers, developers and designers find themselves under pressure to provide more environmentally acceptable products, processes, designs, practices and assets through the ideas of waste minimisation, lower emission and sustainable products. In order to make a substantial environmental improvement and generate business benefit to the society and construction industry stakeholders, environmental issues must be assessed holistically alongside financial, technical and social criteria using appropriate decision making tools.

Life-cycle assessment (LCA) is an emerging environmental decision making tool that enables quantification of environmental burdens and their potential impacts over the whole life-cycle of building assets, products and construction processes in general. Although it has been used in some industrial sectors for more than 20 years, LCA has received little attention in the construction industry sector. Only since the late 1990s has its relevance started to emerge as an environmental tool to aid in the procurement of building assets. This chapter introduces the LCA methods and focuses on the application of LCA in design optimisation as a tool for developing sustainable building assets.

7.2 Life-cycle assessment

Life-cycle assessment as defined by the standard ISO 14040 (Environmental Management – Life-Cycle Assessment – Principles and framework) is a technique for assessing the environmental aspects and potential impacts associated with a product, by:

- Goal defining and scooping of the system under study
- Compiling a life-cycle inventory of relevant inputs and outputs of a product system
- Evaluating the potential environmental impacts associated with those inputs and outputs
- Interpreting the results of the inventory analysis and impact assessment phases in relation to the objectives of the study.

In the LCA context, system boundaries are drawn from cradle to grave to include all burdens and impacts in the life-cycle of a product or process, so that the inputs into the system are considered as primary resources. Figure 7.1 shows graphically the interaction among the LCA phases.

7.2.1 Goal and scope definition

The first step in any analysis must be definition of the system under study. The goal of LCA studies is to quantify the environmental impacts of the building throughout its life in order to minimise them, specifically to quantify raw material use, energy use, emissions to air and water, and solid wastes into an inventory of results. The scope of LCA studies might include the identification of the quantities of the inputs, outputs and associated life-cycle impact assessment. LCA studies might be conducted over the whole life-cycle of a building, process or component. This includes:

- Total life-cycle procurement of the building systems
- Construction and reconfiguration

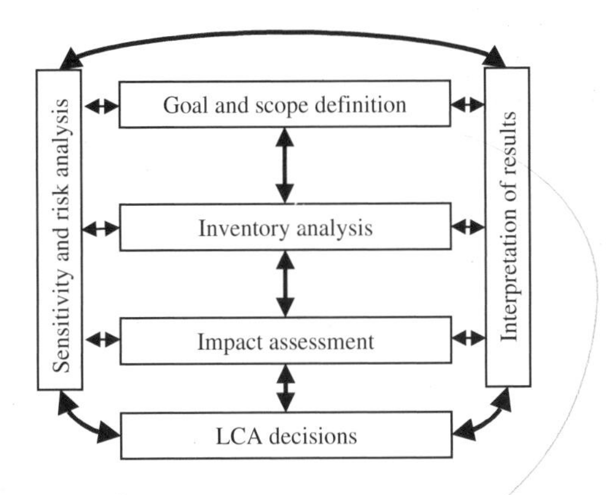

Fig. 7.1 Interaction between LCA stages.

- Operation and maintenance
- Demolition.

7.2.2 Life-cycle inventory

Inventory analysis is a technical data-intensive process of quantifying energy and raw material requirements, atmospheric emissions, waterborne emissions, solid wastes and all other releases for the entire life-cycle of a building asset, component or product. In other words, inventory analysis begins with raw material extraction and continues through to asset disposal stage. Here, material and energy balances are performed and the environmental burdens are quantified. The burdens are defined by resources consumption and emissions to air, water and solid waste. Burdens should be aggregated according to the relative contributions to specific potential environmental effects, such as ozone depletion, land contamination, etc.

7.2.3 Evaluation or impact assessment

Impact assessment is a process to characterise and assess the effects of the resource requirement and environmental loadings identified in the inventory stage. The evaluation should addresses ecological and human impacts, resources depletion, social and all other factors deemed important to the study objectives. The key concept here is to link inventory analysis with impact evaluation by associating resource consumption and effluent releases (environmental burdens) documented in the inventory with potential impacts. Within the impact assessment stage, the impacts may be aggregated further into a single environmental impact function by attaching weights to the impacts to indicate their relative importance. This process is based on subjective value judgements in deciding on the importance of different impacts. In the case of a building the entire impact assessment should include:

- Resource extraction
- Building product manufacture
- Building construction
- Building use/operation/maintenance
- Demolition reuse/recycling/disposal.

The outputs that have potential environmental impacts are categorised into eco-indicators to provide global and some general environmental impacts. Some of the eco-indicators that are produced for the national reuse of waste research programme in the Netherlands include the following impact categories:

- Greenhouse effect
- Ozone depletion
- Heavy metals

- Nutriphication
- Acidification
- Carcinogenesis
- Summer smog
- Winter smog.

7.2.4 *Interpreting the results*

The final phase in the LCA process is interpreting the results of the inventory analysis and impact assessment phases in relation to the objectives of the study. This phase is aimed at identifying the possibilities for improving the performance of the asset. It is also aimed at determining the major stages in the life-cycle of an asset that contribute to the impacts. Since we are dealing with future impacts, sensitivity analysis and risk strategies are developed and final recommendations are made accordingly. LCA studies present an inventory of results of environmental inputs (raw materials and energy) and outputs (air and water emissions and solid waste) from the building assets. These inputs and outputs have potential environmental impacts.

As can be seen from the above process of LCA, its primary objective is to provide a clear picture of the interactions of building assets, processes and practices with the environment, thus serving as a sustainable tool for improving and managing the built environment. Thus, LCA provides a basis for assessing potential improvements in the environmental performance of the asset. This is of particular importance to asset stakeholders, designer and owner because it can suggest ways to modify or design the asset in order to decrease its overall environmental burdens and impacts. The application of LCA in construction is not only restricted to the design phase. LCA has many potential uses, among them the following:

- Urban strategic planning
- Sustainable urban environmental development
- Manufacturers' product and process optimisation
- Construction process optimisation
- Asset operation optimisation
- Asset design optimisation
- Identification of environmental improvement opportunities at all phases of the life-cycle of assets
- Helping decision makers in selecting between alternatives
- Quantifying and evaluating the environmental performance of products and processes
- Assisting in the management of the built environment
- Creating a framework for environmental audits of assets
- Assisting in the management of waste.

Environmental improvements of building assets can only occur when stakeholders are persuaded to consider broader environmental life-cycle thinking and are prepared to integrate it into the decision making process. While a wider acceptance of LCA in the construction industry is still to come, there are indications that the use of LCA is constantly increasing, mainly driven by regulations. The rest of this chapter presents the use of LCA in design optimisation and the integration of LCA and WLCC.

7.3 Life-cycle assessment for design optimisation

One of the most suitable emerging applications of LCA is in the design optimisation of assets over their life-cycle. Environmental considerations have started to be incorporated at early design stages and utilised alongside WLCC measures for selection of design alternatives and optimisation of assets to environmental burdens. Figure 7.2 shows the interaction between design and relevant considerations. During the early stages of projects, LCA decisions are based on guidelines and material rating systems. The LCA decisions at the design stage must be based on extensive use of LCA tools to determine the life-cycle consequences of the design on the environment. LCA decisions at construction stages must be informed by assessment impacts made in the design stage. Decisions at the building operations phase are mainly based on

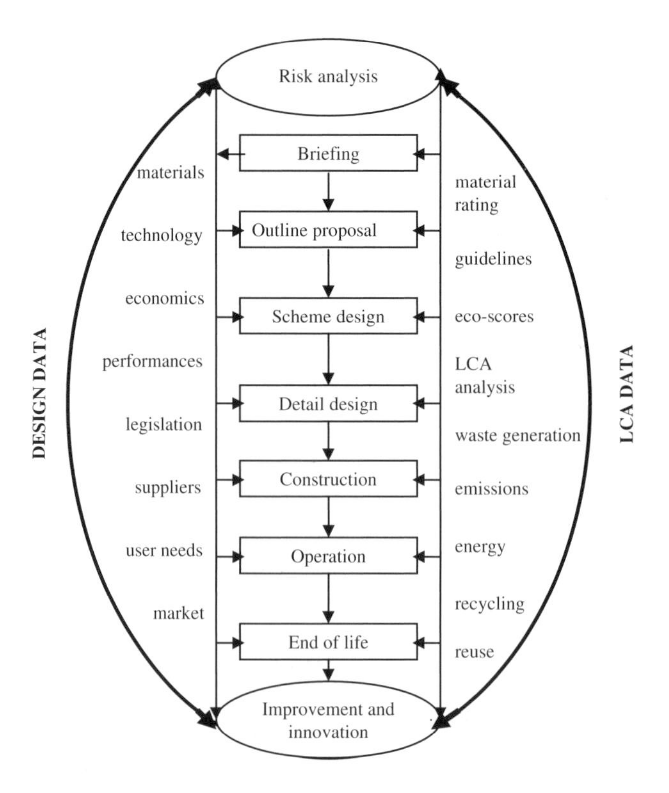

Fig. 7.2

guidelines on how to minimise energy use, water, waste, reuse and recycling. All decisions should be based on guidelines on maintenance and building facilities management. Preferably these guidelines should be based on LCA assessment from the design stage. Guidelines are also required at the end life stage to manage the process of reuse, recycling and disposal of the building components in the most efficient way. The use of LCA in design offers a potential for asset design innovation in the process concept and structure through the selection of the best sustainable technologies and raw materials over the whole life-cycle. Thus, the traditional design practices and processes should be extended to the life-cycles of different technologies and raw materials, all the way from extraction of construction primary resources through to production of components, construction, operation and disposal of assets. This enables a quantitative comparison of different design options of the same set of construction resources as well as an assessment of different raw materials.

It is essential to build a partnership between manufacturers, suppliers and construction stakeholders. For LCA to succeed it is important to include the supply chain within the design LCA process. This enables identification of best manufacturers and suppliers, in term of their product environmental quality and other performance attributes.

The building regulations legislation must be changed so that asset design can comply with environmental emissions as well as all other relevant legislations. Client and end user requirements in terms of specification and environmental performance of the design must be addressed throughout the design process. Once the design parameters and all other associated attributes have been defined, a multi-objective optimisation model for design LCA might be defined. The multi-objective optimisation model might include the following life-cycle processes:

- Supply (raw material extraction, manufacture and transport) of the building systems
- Construction
- Future reconfigurations
- Operation and maintenance over the design life
- Demolition.

The design is then optimised on a number of objectives (extracted from client and end user requirements), defined from environmental burdens, WLC, socio-economic considerations and economic, technical and functional life of asset, subject to the constraints on material and energy balances, durability, functionality, legislation, and all other perceived relevant constraints. The design optimisation process will result in a number of design alternatives, enabling a quantitative evaluation of options for environmental, technical, functional, economic and any other aspects of design improvements. The design procedure is dynamic with a continuous interaction and exchange of information within the project design team and other stakeholders to explore systematically the possibility for continuous improvements. This whole procedure must be assisted by risk assessment of all key decisions within the life-cycle of the asset under design. In this respect, LCA can provide a powerful tool for the design of cleaner assets that are environmentally sound and economically profitable.

7.4 LCA tools

To assess the environmental impacts of buildings, several tools are being developed. Some of these tools rely directly on LCA methodologies while others use environmental impact with a simpler approach based on qualitative and quantitative methods. Most of the existing LCA tools cover only some aspect of the life-cycle, like embodied energy and CO_2 emission and computation of operating energy of building assets. This section gives a short overview about some of the existing methods concerning life-cycle assessment for buildings and their components. LCAs are categorised into five major classes (Centre for Design 2001).

(1) *Tools* that focus only on materials, components and processes embody energy and environmental impacts. These tools are mainly used in selecting materials, while also allowing material producers to optimise production processes (e.g. BEES). The BEES tool is based on methodology for selecting building products that achieve the most appropriate balance between environmental and economic performance based on the decision maker's values. The aim of the BEES tool is to generate environmental performance scores for building products in the USA. The LCA scores are combined with economic scores to help decision makers select cost-effective, environmentally friendly building products (BEES manual). Readers can access the website of the BEES for more information: www.bfrl.nist.gov/oae/software/bees.html

(2) *Design tools*, which use LCA as a basis but are simplified to single indicator points or to an aggregation of impacts to building component levels. An example of such tool is the ENVEST modelling tool. ENVEST is a software tool for estimating the life-cycle environmental impact of commercial buildings, schools and hospitals at the building inception stage. The tool is being developed by BRE UK, which claims that it provides a holistic approach to the design process by helping to optimise the form of the building for the least environmental impact over the building life-cycle. ENVEST considers the environmental impacts of both the materials used during construction and the energy consumed over the building's life. Data on 12 environmental impact categories is multiplied by a subjective weight for impact category and combined to produce an overall output as an eco-point score. The system also allows for WLCC computation. Possibly the advantage of this tool is its ability to allow designers to identify those aspects of the building which have the greatest influence on its overall environmental impact and cost.

Readers can access the website of the ENVEST for more information: www.bre.co.uk/service.jsp

(3) *Integrated LCA/CAD tools* have the capability to read material and component information from CAD drawings and perform an environmental impact analysis. An example of such tools is the LCAid™. LCAid™ is a Windows-based computer modelling environment for LCA developed by New South Wales Department of Public Works and Services, Australia. The software objective is to assist designers and practitioners in LCA and design improvements. The model allows design ideas and options to be tested throughout the design stages. It has the advantage of being integrated with other 3D CAD tools like Ecotect. LCAid™ uses quantities of asset components in addition to operational energy and water consumption to model/compute environmental burdens. These are calculated over 11 eco-indicators covering:
   - Atmospherics
   - Resources
   - Pollutants.

   The environmental impacts within each eco-indicator are separated into four stages: construction, operation, maintenance and demolition. The advantage of LCAid™ as a tool is its ability to allow the user to find where design improvements might be made and assist in the environmental decision making process, allowing the overall environmental impact of assets to be reduced. Readers can access the website of the LCAid™ for more information: www.projectweb.gov.com.au/dataweb/lcaid/

(4) *Green product guides and checklists* are qualitative guides to products and issues which need to be considered (e.g. BREEAM). BREEAM (Building Research Establishment Environmental Assessment Method) is a UK tool. There are several versions of it, which cover several types of buildings including offices, new superstores and supermarkets, and new homes. For example, the BREEAM for new homes method seeks to minimise the adverse effects of new homes on the global and local environment while promoting a healthy indoor environment.
   The main impacts of building identified in BREEAM are:
   - Global atmospheric pollution (greenhouse effect, acid rain, ozone depletion)
   - The local outdoor environment and depletion of resources
   - The health, comfort and safety of occupants
   - Local outdoor climate (e.g. changes in wind loading).

(5) *Building assessment schemes* are used to assess whether a building is performing adequately, sometimes allocating star ratings. Mostly they are applied post-construction; however clients will often specify a design rating for the building preconstruction (e.g. GBTool). The GBTool software has been developed as part of the Green Building Challenge

process, an international effort to establish a common language for describing 'green buildings', which now includes teams from 20 countries. The software has been developed by Natural Resources Canada on behalf of the GBC group of countries. The tool is used to assess predicted or 'potential' performance of a building before occupancy. It is not intended to assess performance during operational conditions. It is applicable to offices and multi-unit residential and school buildings. Readers can access the website of the GBTool for more information: http://greenbuilding.ca

7.5 Environmental life-cycle cost

In Chapters 2 and 3 we discussed how WLCC analysis influences the whole cycle decisions and provides possible explanations of the relationship between capital costs, WLC costs and whole life key decisions. WLCC studies can contribute to an asset's cost reduction by identifying high environmental cost aspects of the design. The combination of rising waste, environmental costs, emissions and scarcity of resources, etc. has created an awareness of and interest in the eco-costs of asset ownership. Stakeholders must give importance to eco-costs throughout the life-cycle of assets. While there are several models for analysis design for environmental burdens, there is a lack of mathematical models for addressing the eco-costs of the environmental burdens caused by the procurement and operation of assets in their entire life-cycle.

As illustrated in the previous section, LCA provides a framework for measuring and benchmarking indicators that demonstrates the environmental performance of procured assets to the stakeholders and society in general. But without incorporating eco-costs into LCA analysis, well-informed decisions on assisting environmental performance and investments cannot be made. Mechanisms that carefully track environmental costs and impacts will greatly help building stakeholders in meeting the performance requirements set by all relevant legislation. Thus, it is prerequisite that cost analysis must include eco-costs into WLCC studies of building assets. Eco-costs are both direct and indirect costs of the LCA, impacts caused by the building asset, product, process, etc. in its entire life-cycle. WLCC models that incorporate eco-costs can assist in identifying eco-design alternatives and can help in reducing total ownership cost through eco-friendly decision alternatives throughout the life-cycle of building facilities.

A building asset eco-cost model might include the following cost breakdown structure (CBS):

- Cost of controlling atmospheric emissions
- Cost of resources (i.e. energy and water consumptions) used in the extraction and production of product
- Cost of waste disposal

- Cost of waste treatment including solid and other wastes
- Cost of eco-taxes
- Cost of pollution rehabilitation measures
- Cost of environmental management.

The above costs must be attributed to the process of the building life-cycle as described in the following subsections.

7.5.1 Eco-costs at resource extraction stage

The environmental costs associated with resources extraction might be considered as non-recurrent costs. The costs occur only at the stage of the extraction process as long as no environmental rehabilitation measures are necessary to mitigate the impacts created by the extraction process over a long period of time. If the latter is the case these costs should then be discounted and relating cash flow projected over the required period.

7.5.2 Eco-costs at building product manufacturing stage

The eco-costs relating to building production include costs of energy, packaging, transport, waste and emissions. Product manufacture bears these costs. But usually these costs are passed on through the supply chain system. In one way or another the end user will bear the majority of these costs as well as society in general. It is possible to minimise these costs through innovation in manufacturing processes, selection of eco-friendly raw materials, and production of products that are less energy intensive and produce less waste and emissions. Eco-friendly building products cost more than traditional products; however, with improvement in the manufacturing process these costs will be reduced. It is possible in the foreseeable future that the majority of building products will be labelled according to their eco-impacts. Those products that are less eco-friendly should carry higher taxes, and this may lead to equilibrium of costs between the two products.

7.5.3 Eco-costs at building construction stage

The eco-cost at the construction stage is mainly due to the resources expanded in the building process, transport and waste. These costs can be minimised through sound construction planning and buildability to eliminate all type of waste. This requires the development of sustainable construction methods that use a minimum of energy, generate less waste and have minimum environmental impacts. The eco-costs of civil engineering projects are higher than the eco-costs of building facilities. Civil engineering projects by their nature require a great amount of resources, but the environmental impacts from these types of projects tend to stop by the end of the construction period. The environmental impact of building projects is mainly in the operation and use stage.

7.5.4 Eco-costs at building use/operation/maintenance stage

The majority of the eco-costs occur at this stage of the asset's life-cycle. The eco-costs here are mainly due to energy and water consumption, solid wastes, eco-taxes and all outputs (emissions to air, water, land and waste). It is possible to optimise all of these eco-costs through the application of LCA and WLCC at the design stage by selecting the best eco-friendly design alternatives. A sound facility management at the operation stage of the asset is also a prerequisite.

7.5.5 Eco-costs at demolition stage

Building waste from demolition can be regarded as potential raw material for production of new materials. The options available for dealing with demolition waste are land fill, incineration, recycling, remanufacturing and reuse. The eco-costs of each option depend largely on the methodology and processes used in the demolition. Eco-costs here might be attributed to taxes and emissions.

7.5.6 Eco-costs at reuse/recycling/disposal stage

Recycling building materials can considerably reduce the use of energy and extraction of natural resources and reduce the use of land for landfill. The end-of-life scenarios identified by industrial ecologists include:

- Reuse
- Repair
- Reconditioning
- Recycling of materials
- Disposal.

The first three options are the most eco-friendly and cost effective. In the disposal option, the main part of the total embodied energy and natural resources used to produce the product is lost. Also the disposal eco-costs can be very high. Recycling can bring a considerable part of the embodied energy back into use. The scope for recycling building materials/components depends to a high degree on how buildings are designed. Design for disassembly and recycling can serve as a stimulus for increased future recycling and eco-cost reduction.

In order to assess the eco-costs potential of each option, data on all outputs, solid wastes, eco-taxes, and disposal, reuse and recycling techniques and their energy requirement must be assessed and compared. Thus, the eco-costs here depend largely on the option being adopted. It is expected that a combination of options might be used at the end of building asset life.

Each of the above eco-cost categories may have a defined relationship with life-cycle stages of asset developments. That is why the elements of eco-costs illustrated above must be added to the WLC costs at the relevant stage and to the relevant cost element. There are several options on how this can be done. Among these options is that eco-costs can be estimated and added to the WLCC elemental costs in Chapter 8. Eco-costs can be estimated as a:

(1) Percentage of the capital cost of each element
(2) Cost per unit
(3) Percentage of the total cost
(4) Product-related cost
(5) Combination of the above.

Among the above options, probably the product-based approach is the most effective way for identifying eco-design alternative products and solutions. In this approach the eco-costs are directly related to the product. This will force manufacturers and suppliers to understand the consequences of their products and it may accelerate the development of eco-friendly construction materials.

7.6 Case study

The methodology followed in the case study is based on an approach employed for analysing similar buildings in Australia. The methodology is reported in great detail in Centre for Design (2001).

7.6.1 Goal

The goal of the LCA on the office building is to quantify the impacts of the building throughout its life in order to minimise them, specifically to quantify raw material use, energy use, emissions to air and water, and solid wastes into an inventory of results.

7.6.2 Scope

The areas covered by the life-cycle assessment of the house include:

- Procurement (raw material extraction)
- Manufacture and transport of the building systems
- Construction and reconfiguration
- Operation and maintenance for a 60-year design life.

7.6.3 Functional unit

The functional unit is the provision of a small office building for 60 years. This LCA further split the functional unit per life-cycle stage, as summarised here.

General: Each stage of the life-cycle assessment is based on a functional unit, to which all the results are referenced. These are detailed below.

Total life-cycle: The functional unit for the total life-cycle is the sum of the functional units for procurement, construction and reconfiguration, operation and maintenance stages.

Procurement: The functional unit for the procurement stage is the raw material extraction, processing and transport of the major building materials to the house site.

Construction and reconfiguration: The functional unit for this stage is the - construction of the house (i.e. the building systems considered in procurement) and the reconfiguration of the house to suit different occupant styles.

Operation and maintenance: The functional unit for this stage is the operation and maintenance of the office over its 60-year design life.

7.6.4 System inventory analysis

Building system procurement: This includes extraction of raw materials, manufacture of the products and systems and transport. All major processes, which are required to procure the building systems such as the production of energy and intermediate transport, should be included.
Office construction and reconfiguration: This Might include:

- *Waste management*: waste to landfill, waste to reuse or recycling
- *Energy use*: diesel and gas fuel for construction
- *Water use*: potable town water
- *Raw materials*: additional systems added in reconfiguration.

Operation and maintenance: The issues that could be considered over the 60-year design life of the house include:

- *Energy use*: heating and hot water, lighting, kitchen facilities and power
- *Waste management*: waste to landfill, waste to recycling and combustible waste
- *Water use*: tap water and water collected on site
- *Maintenance*: replacement of systems as required, waste produced in maintenance.

7.6.5 Inventory data

The inventory data is presented in Table 7.1. BRE ENVEST software is used to extract the embodied eco-points.

Table 7.1 Inventory data.

Inventory data for a small office building consisting of two floors and located in the north-west of England	Embodied eco-points
Floors	
Ground floor: in situ reinforced concrete (200 mm) and expanded polystyrene for insulation	709
Upper floors: precast slabs (150 mm)	495
Walls	
External walls, brick and block	567
Internal walls, aerated concrete (100 mm)	54
Windows	
Windows UPVC double glazed	119
Roofs	
Structure, timber	676
Covering, plain concrete tiles	568
Finishes	
Floor, wool carpet	350
Wall, emulsion paint	121
Ceiling, plasterboard	71
Structures	
Foundation, strip 450 × 300	130
Services	
Heating	2507
Lighting	3195
Ventilation	338
Water (including water heating)	318
Lifts	826
Office equipment	2307
Catering	974
Other	423
Total	14748

7.6.6 Results

The results of the LCA analysis are shown in Fig. 7.3. The figure shows the distribution of embodied eco-points for the fabric and operation of the office. A comparison between operational and embodied is also presented. The figure shows that operational eco-points are extremely high. This suggests the operational strategy adopted for this office is inadequate. In this scenario analysts must go back and modify those input parameters that have a high effect on operational eco-points.

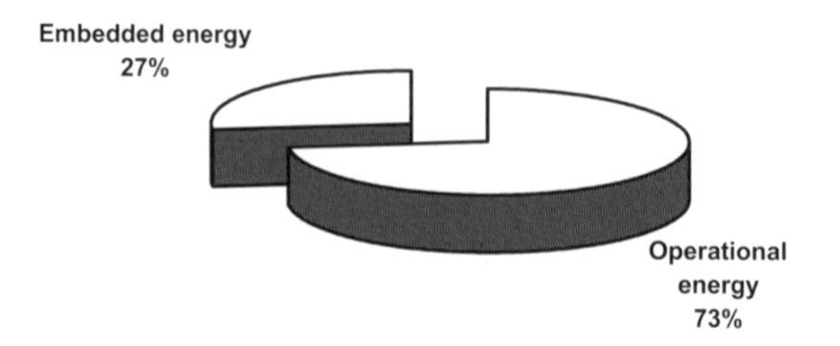

Operational versus embodied energy

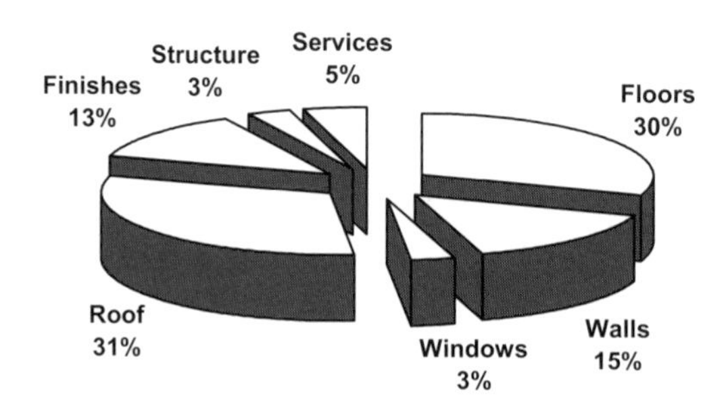

Embodied eco-distribution

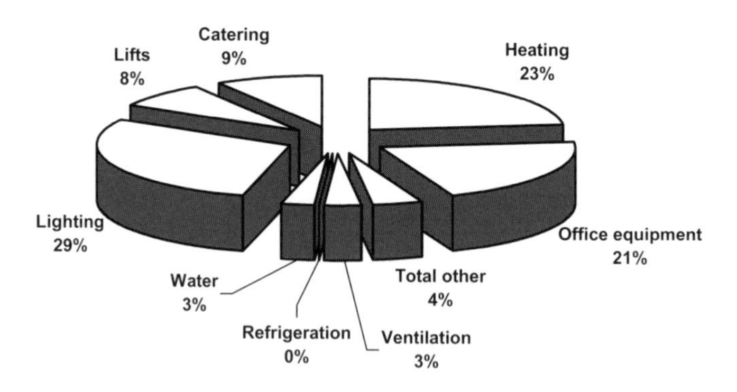

Operational eco-distribution

Fig. 7.3 Results of LCA analysis.

7.7 Summary

Utilising a life-cycle assessment approach to quantify the environmental liability of a building design and the potential impacts over the whole life-cycle of a building should form an important facet in the overall WLCC approach. The very fact that LCA has received little attention in the construction industry sector over past years provides the ideal opportunity to develop models which provide far more accurate information, and help clients and designers specify more sustainable design solutions. Substantial environmental improvements that can be obtained from an integrated WLCC/LCA approach can create business benefits not only to the client and society at large but also to the construction industry stakeholders who are now under increasing pressure to deliver quality sustainable design.

Reference

Centre for Design (2001) *LCA Tools, Data and Application in the Building and Construction Industry*. RMIT University, Australia.

8 Whole Life-cycle Cost Planning at the Design Stage

8.1 Introduction

Decisions regarding the selection of construction technology and construction materials are no longer based solely on technical and economical aspects but are becoming increasingly influenced by WLCC and environmental considerations. The ability to influence the outcomes of total whole life ownership is greatest during the design phase as the types of material specified, the quality of the design and the contracting method chosen impact directly upon operation and maintenance costs. Also, the procurement methods used may have implications and a great influence on whole life-cycle costs. Operating, maintenance and rehabilitation costs of new and existing facilities amount to more than 80% of total life-cycle costs. It is well documented that the majority of decisions about these costs are predetermined at the design stage. The opportunities to modify or influence these decisions diminish as projects progress through their natural process of development. Hence, risks and consequences of these decisions on the total cost of ownership of assets must be identified and planned for. That is why it is important to establish a mechanism at the design stage that brings together the WLCC, service life, environmental life-cycle assessment, and risk associated with decisions taken at this stage.

The last two chapters have addressed the issues of design environmental life-cycle assessment and design service life estimation. This chapter and the next will deal with whole life-cycle cost planning and whole life costing risk and risk responses at design stage.

8.2 Design whole life-cycle cost planning

Whole life-cycle cost planning at the design stage must provide an exhaustive accounting of all resources required for the acquisition, operation, maintenance and disposal of assets. The WLC budget estimate must be comprehensive and structured to account and identify all cost elements. It includes total cost of ownership over building assets life-cycle (i.e. economical, technical and functional lives). The WLC budget must include costs for feasibility, scheme, final design, implementation, operation, disposal, environmental and risk reserves. To achieve this aspiration, WLC planning at the design stage requires a formal framework that considers all aspects of building assets, a framework that permits data and information transfer between design stages and provides benchmarks and guidance within which whole life cost control may be exercised. It could also assist project stakeholders to identify critical aspects of the asset so that effort can be expanded on these areas.

This is a challenge that calls for rethinking the way we used to cost projects at the design stage. This chapter provides the foundation for such a challenge and addresses aspects of whole life cost estimation throughout the design stages. The chapter does not address or illustrate the issues of building rates and modifying these rates to regional and other variations.

8.3 An integrated framework for WLC budget estimation

Figure 8.1 shows a generic step-by-step framework for developing a WLC budget estimate. Notice that the entire process is continuous. Most of the steps within this framework are iterative and may occur at various stages within the overall process. The framework consists of the following iterative steps.

Understanding client objectives

Before any WLCC estimate is carried out an analyst should identify and understand the clients' requirements, objectives and needs. The assessment of a possible operational environment for identifying risks and deficiencies in

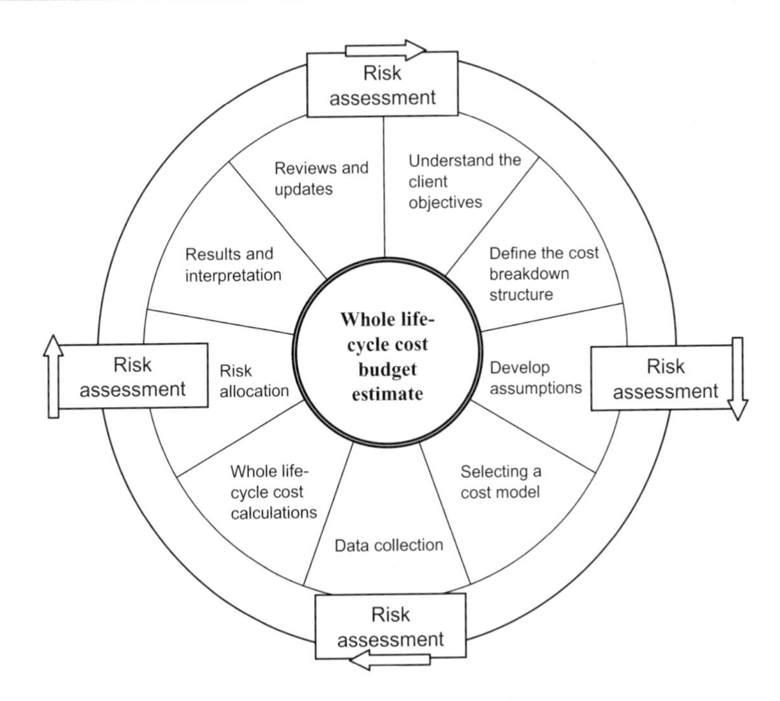

Fig. 8.1 An integrated framework for WLCC budget estimation.

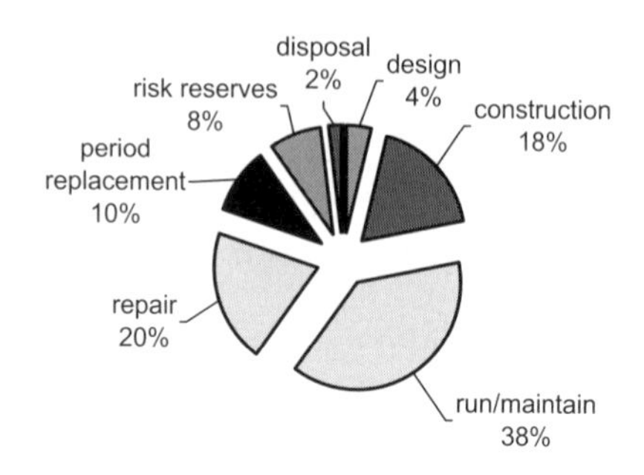

Fig. 8.2 Benchmarking cost of total ownership.

the clients' objectives is also required. It is important at this early outset to establish WLC budget benchmarks, like those in Fig. 8.2, to which the actual WLC budget estimate can be compared and controlled. Chapter 2 has further details on client requirements.

Define cost breakdown structure (CBS)

We suggest that a WLC budget estimate should be progressively defined and broken down into all its life-cycle constituent components. The finer the granulate level of the breakdown the more accurate the whole life cost planning will be. The CBS should reflect client objectives and needs. For this purpose BCIS elemental cost analysis can be modified (by adding WLCC and LCA data to each major cost element) and used to define a WLC CBS. A structured and consistent CBS is essential to capture all whole life costs, to communicate information among stakeholders, and to provide a baseline for WLC cost control. An effective CBS should be product focused and include the service life of the product, operation cost, maintenance costs, components replacement frequencies, reserve risks and risk of failure information.

Develop WLCC assumptions

Every aspect of cost modelling and WLC budget estimation is based on assumptions drawn at the time of performing the modelling process. Properly defined WLCC assumptions are the most significant tasks in the WLC budget estimation process. They are important in costing the scope of the proposed investment and will help stakeholders in understanding what it is being included and excluded from the current WLC budget estimate so that future comparisons can be carried out objectively. A WLC budget based on inaccurate assumptions may lead to inappropriate decisions and a high risk of investment failure. All WLCC assumptions should be documented in a logbook from the outset so that if there are any changes in any circumstances the analyst can refer to this documentation. Assumptions should also reflect client needs and aspirations and preferably be developed in coordination with the project participant. A detailed list of possible assumptions for each of the design stages is included in the following sections and in Chapter 2.

Selecting a cost modelling method

The selection of a WLC cost estimation methodology is tied up to the life-cycle of the assets development process. The methodology will largely depend upon requirements being identified, data availability, project complexity and analyst skills. It is expected that a combination of methodologies might be used to develop WLC budgets. Regardless of the methodology utilised, assumptions built around the costing method must be clearly understood so that the results can be interpreted accurately. A process or a mechanism that enables decision makers to appreciate, understand and utilise cost model outputs confidently should underpin successful selection of cost models. For an in-depth detail of WLCC models readers should refer to Chapter 4.

Data collection

The world we live in is a complex dynamic place. Information changes all the time. If we believe that we are in a world economy that is knowledge driven, then data in all its forms is essential to complete an exhaustive and comprehensive evaluation of WLC budgets.

Data collection and interpretation is a continuous process that occurs iteratively at each stage of the WLC budget development process. Data should be understood and modified if necessary, before it is used. While WLC cost data is difficult, time-consuming and very expensive to collect, this should not be used as an excuse for not conducting a WLC budget analysis. With minimal data and some experience one can develop very elaborate models. WLCC data requirements may appear unclear in the early stages, but data requirements often evolve as projects pass through their natural life-cycle stages. Sources of data and the data collection process are explained in detail in Chapter 2.

WLC budget estimate computation

Having developed the assumptions and collected the necessary data, now WLC budget estimate can proceed. All input information and formulas should be checked for errors and conformity to the original assumptions. WLCC computation can be carried out using WLC standard sheets, Excel models or stand-alone programmes; the choice largely depends on task complexity and user sophistication.

Budgeting for WLC risks

Chapter 5 deals with risk theories and methods for estimating risks. Budgeting for WLCC risks includes the establishment of contingency reserves for the design, construction and operation of building assets. Part of this is the determination of risk influencing factors and carrying out sensitivity and risk analyses. WLCC risks can be accounted for in each of the budget assumptions using probability distributions (minimum, most likely, and optimistic values).

Distributions can be easily defined for all key WLC budget elements. The results can be presented as cumulative probability distribution curves at different percentile intervals. These curves can then be used to estimate and allocate WLCC risk contingencies to achieve a desired level of confidence that a WLC budget would not be exceeded.

Results interpretation

The quality, quantity and type of results generated from a WLC budget estimate analysis depend largely on the complexity of the project being budgeted for. The results can vary from a simple table to very sophisticated simulation and probability graphs supported by statistical information. Whatever, a standard format should be adopted so that the results interpretation can be consistent between all project participants. All results should be validated for consistency and accuracy. All WLC budget estimates that are thought to have a significant risk attached to them should be highlighted and brought to the attention of decision makers.

WLC budget review and updates

Estimating a WLC budget spans over long periods of time. Prices, inflation, taxes, rates, client requirements, etc. may change over this time. Thus, WLCC results must be updated periodically or whenever there is a change in the initial assumptions. This will provide decision makers with real-time information so that risk mitigation strategies can be acted upon.

8.4 Benchmarking WLC budgets

Chapter 3 explained the concept of benchmarking in WLCC. Benchmarking WLC budgets should be a critical component of the information needed to make decisions in order to implement proposed clients' requirements and to evaluate the success of decisions throughout the life-cycle of building facilities. A WLC budget benchmark should be used to influence the decisions to proceed with the development, based on the total financial resource commitment shown in Figs 8.2 and 8.3. The type of information presented in this figure and in WLC budget estimates assists decision makers to determine the appropriate scope or size of the facility they would like to develop. Thus, underestimating WLC budgets will prevent stakeholders from making inappropriate investment decisions, although inflating WLC budgets may result in the investment being deemed unviable with the risk of loss of the investment opportunity.

The margins of error or percentile range of a WLC budget made early in a project development are relatively large, compared to those made in the later stages. This is mainly due to the fact that as information becomes readily available, cost analysts are able to utilise project details to improve WLC

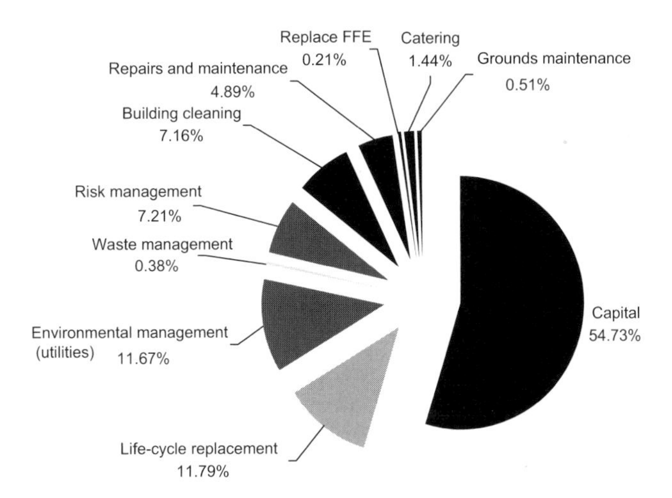

Fig. 8.3 Whole life cost: where is the value? PFI school whole life costs 30 years (Bartlett 2003). (Reproduced with kind permission of the author.)

budget estimates. The range of cost methods that could be used to model WLC budgets through the life-cycle of facilities is shown in Fig. 8.4. The cost methods vary from broad range estimates based on £/m^2 to detailed site work measurement. Figure 8.4 shows graphically the relationship between project life-cycle stages, the margin of error and cost methods that may be used to derive WLC budgets. The level of accuracy shown in the graph is mainly for illustration; it is well known that the actual WLC budget exceeds the budget estimate. Underestimates of budget costs are the norm rather than the exception. The inaccuracy of the process is reflected in the costing method used to derive the WLC budget estimate.

As explained previously, WLC budget planning is a progressive process. Time, cost and information uncertainties have a significant effect on building

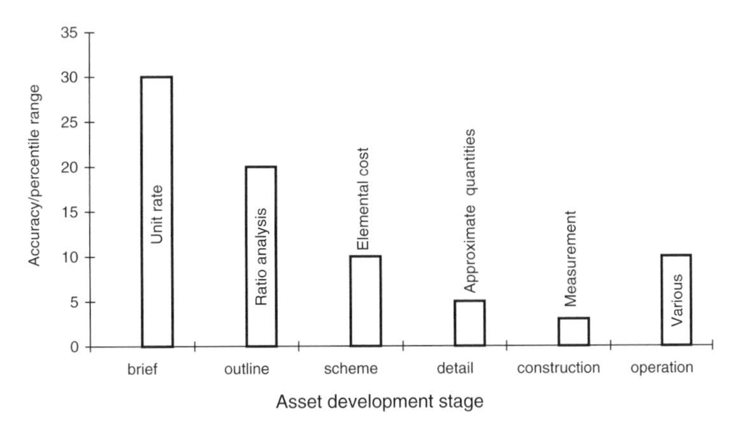

Fig. 8.4 Relationship between accuracy of whole life-cycle estimate, stage of development and cost estimation methods.

assets' budget accuracy. This makes it almost impossible to derive a WLC budget with a high certainty. That is why a rethink of the way we estimate and interpret costs throughout the life-cycle of building facilities is overdue. We need to think in terms of cost movement and trend rather than accuracy of budget estimation. If we know the cost trends and their influencing variables then we can develop strategies to deal with most risk eventualities. An approach based on this concept will allow stakeholders to allocate target costs as percentages for each cycle of an asset and distribute the cost over the life span of the asset according to these targets. Thus, the process is reduced to cost control movement and trends within the predefined targets. An example of benchmarking in WLC budgets is to allocate the total budget as a percentage among the CBS of the asset. Figure 8.3 presents the issues graphically. Such benchmarks could be extracted from best-performing building assets and used for budget distribution among building elements, and they could be used for optimising or trade-offs between operation and construction costs. Allocation of WLC budgets in this manner will assist asset stakeholders to utilise value-engineering techniques to identify unnecessary costs within each phase of the life-cycle of projects, thus ensuring the best value for money concept.

8.5 Whole life cost planning

8.5.1 Whole life-cycle cost planning at briefing: inception stage

Clients and their representatives are required to prepare a general outline of requirements of a proposed development so that early WLC budget advice can be provided. It is normal practice at the design stage to estimate facilities costs based on building components that make up the entire facility. The detail of components or building elements used in the estimate largely depends on the stage of the design process at which the WLC estimate is being made. In the very early stages of building assets development, whole life-cycle cost planning should be based on general information as shown in Table 8.1. Traditionally, cost advice at this stage is based on coarse cost information such as £/m^2, without the inclusion of the operational costs. The authors advocate that if the concept of total ownership is to be taken seriously, then all future costs should be taken into account at this stage so that clients are able to appreciate what they are committing themselves to. A sound plan for future actions and decisions is also required here. Table 8.1 shows an example of a whole life-cycle budget cost estimate. The cost estimate of each element is based on the −20% to +30% percentile range estimate discussed in Chapter 6.

Inception WLC budget assumptions might include:

- Building cost
- Land cost
- Cost of finance
- Site demolition costs if required
- All fees, including professional and legal

- Taxes and insurances
- Cost fitting and furnishing
- Operational costs
- Disposal costs
- Environmental costs.

Initial analysis and projections of the costs relating to the above assumptions over the life-cycle of building assets could assist clients in their decision to go ahead with a project or otherwise without incurring further costs for the next development stages. The outline requirements of the project are first established and budgets for the above are determined based on techniques like cost per number of users, cost per m^2 or other preferred methods. Usually information for this purpose is acquired from past projects and expert judgement. Then the risk associated with each of the above budget elements is evaluated and a cost variation estimate is determined based on the level of anticipated risk. Since budget projection is over a long period, where time and cost are uncertain, a simulation and sensitivity analysis ought to be carried out to generate several estimates with associated confidence levels. Financial information with other qualitative benefit factors should then be used to inform stakeholders if the building asset provides good value for money. The process should demonstrate with high certainty that the benefits to be derived from the building asset significantly exceed its lifetime costs. Social and financial benefits are related to the function that a facility performs during its life span. That is why it is very important to acquire the right facility based on clearly stated clients' objectives. Any aspect that does not contribute to the clients'

Table 8.1 WLCC budget estimation at inception stage.

Budget item	Cost £/m^2			Annualised £/m^2 @ 6% rate		
	Min. @ 80% variation	Most likely	Max. @ 130% variation	Min. @ 80% variation	Most likely	Max. @ 130% variation
Initial capital cost	2400	3000	3900	206	258	335
Operation and maintenance costs						
As ratio to capital cost 1:2 per year	1200	1500	1950	1200	1500	1950
Replacement costs @ year 10	400	500	650	18	22	29
Including inflation @ year 15	560	700	910	17	21	27
@ year 20	800	1000	1300	18	22	29
WLC risk reserves 20% of capital cost	480	600	780	41	51	53
Resale value	160	200	260	14	17	22
Total				1514	1891	2445

objectives or the functioning, durability and running of the proposed facility has no value and is an unnecessary add-on cost. Thus, WLC budgets at this stage should be based on clearly-defined client requirements, so that unnecessary expenditures are eliminated and an optimum balance between cost, time and quality is obtained.

To demonstrate the above theoretical concepts let us now consider an example for developing a WLC budget for a wet/dry multipurpose sport centre for a local authority client. The client has several options for acquiring this project. Among the feasible options is to use a PFI to lease the facility over 25 years. Another option is an own build, but the cost of the project is financed at an interest rate of 6% with the possibility that this may rise to 8% over the life span of the project. Using the PFI option would cost the local authority £2000/m^2 per year over the next 25 years, all-inclusive. The centre consists of 5000 m^2 gross floor area and a 25 m long swimming pool. The anticipated service life of the project is 25 years. The centre is located in the north-west of England on a brown field site with no soil contamination but requires demolition and recycling of the existing site materials. The interest rates that should be used for WLC budget-planning computation should be based on the anticipated return which the investors require in order to make a profit from the investment. It is advisable that a sensitivity analysis for different rates should be considered so that the stakeholders understand the implication of their decisions. Table 8.1 shows the cost assumptions and calculation for the own build option. The third column shows the annualised cost using a yearly periodic payment schedule. The formula for computing these values is included in Chapter 4. This is necessary so that it enables us to compare the annual leasing cost with the build option cost. The most likely values for each budget element might be extracted from past projects. Range values are based on a percentage of the most likely value. In this scenario ranges are based on historical data from typical projects. Refer to Chapter 6 for a more elaborate method for defining ranges based on the anticipated risk associated with each budget item.

In Table 8.1 initial capital cost is converted to an annualised value using formulae in Chapter 4 or conversion tables. Since operation and maintenance costs are expressed as yearly values, conversion to annualised values is not required. Replacement costs are converted to present worth values then expressed as annualised amounts. The costs of risk reserves and resale value are also converted to annualised figures. From the figures shown in Table 8.3, column 3, the clients may decide that it would be more economical to build if anticipated risk and cost variation is less than 30% of the most likely costs.

8.5.2 *Whole life-cycle cost planning at briefing: feasibility stage*

If the initial analysis proves to be viable and decisions are made to proceed with the project, clients then consider in some detail the project outline requirements and build a team to further develop the initial concept ideas of the project. This includes the consideration of environmental, functional, technical and financial requirements. WLCC information is required in order to establish the overall viability of the building asset over its life span. One of the major decisions that will influence the project outcome at this stage is the selection of who will design, manage, construct and operate the asset. This will have a direct effect on the design, quality, functionality and post-occupancy costs of the built asset. The procurement of built environment assets can be viewed as a problem-solving process that requires effective team working between members with separate responsibility and skills, to achieve the best results. In the case of the whole life-cycle procurement process the team should be involved at the early stages of the asset development so that design, construction and the operation phases are tightly linked, with the aim of offering best value and continual innovation for improving the performance of the whole team.

The team is required to carry out investigations into user/client requirements, site conditions, planning, design, operation costs, construction cost, legal aspects, market conditions, technological aspects and all operational matters relating to the asset. Information generated from these investigations is essential for informing the decisions that are taken at this stage and providing clients with an appraisal and recommendation for the best option, based on WLCC aspects and anticipated risks. The outcome of the feasibility planning will determine the form in which the project is to proceed, ensuring that it is functionally, technologically, operationally, environmentally and financially viable. Projects become viable when the team can find the right balance between all these factors. In establishing the balance the team should bear in mind that every £1 saved from the operational cost through good design is equivalent to a £10–12 reduction in capital costs (Kirk & Dell'Isola 1996). Hence it is important to recognise that the effect of savings on operation costs is more important than savings in projects' capital costs. The potential of saving using WLCC at the feasibility stage is very significant, through a clear definition of client objectives. It is important to elicit and determine clients' cost priorities and establish procedures and metrics for defining these costs. WLC budget cost boundaries must be established and communicated to the team members.

The assumptions required for the second phase of WLC budget planning might include:

(1) *Initial project brief*
- building client/user
- size, floor area, space, number of floors, shape, location, etc.
- aesthetic aspects
- functionality and quality
- details of the site
- planning issues.

(2) *Cost information*
- construction cost prices from past projects
- prices for internal finishes from similar projects
- running costs for similar projects

- maintenance requirements and costs of similar facility
- prices for external works
- all fees associated with the development and operation of the project
- preliminaries
- whole life contingencies (including design, price, construction, operation and service life risks).

(3) *Risk influencing factors*
- inflation
- interests
- market condition
- user
- technological
- organisational (client and stakeholders)
- availability of resources
- time constraints.

(4) *Service life*
- functional service life of components and facility
- technical service life of components and facility
- economical service life of components and facility.

To demonstrate the concept of WLC budget estimates at the feasibility stage, let us consider an example of building a hospital general ward. The ward is intended to cater for 100 beds and associated accommodation to run the ward. The total floor area of the ward is 3150 m^2. Some of the WLC budget estimates shown in Table 8.2 are based on historical data from past similar projects. Data for replacement and decommissioning is hypothetical and used for illustration purposes. The data is modified to take into account all risk aspects and adjustment for inflation, location, market conditions, etc. The method of modification is well documented in other sources (Flanagan & Tate 1997). Annual operational and maintenance costs are based on current costs of similar facilities. The frequency of alteration and replacement is also extracted from similar projects and is mainly related to equipment and flooring changes. Triangular distribution is used to model the estimated budget costs. Other forms of distribution and estimation range might also be used; see Chapter 4 for more detail. This information can then be used to simulate and find the most probable whole life-cycle cost budget. Preliminaries and risk reserves are expressed as an approximate percentage of the capital budget.

8.5.3 Whole life-cycle cost planning at outline stage

The overall viability of the project is established over the last two phases. The team has to develop further the brief and carry out investigations as necessary to reach WLCC decisions. An overall sketch design with regard to layout and construction should be developed. Alternative design forms and construction methods should be considered along with their operational strategies and

Table 8.2 WLCC budget estimation at feasibility stage.

Asset life-cycle: 25 years Discount rate: 8%	Estimated cost £/m²			WLC annualised @ 8% discount rate		
Budget item	Low @90%	Base	High @120%	Low @90%	Base	High @120%
Ward initial cost	945	1050	1260	88	98	118
Fees @ 15% of the ward cost	141	157	188	14	15	18
Equipment cost @ 40% of the ward cost	378	420	504	35	39	74
Subtotal	1464	1627	1952	137	152	210
Maintenance per year:						
Building	5.4	6	7.2	5.4	6	.2
Engineering	9	10	12	9	10	12
Grounds	1.4	1.5	1.8	1.4	1.5	1.8
Operation per year:						
Cleaning	18	20	24	18	20	24
Staffing (porters)	2.7	3	3.6	2.7	3	3.6
Security	0.9	1	1.2	0.9	1	1.2
Laundry	3.6	4	4.8	3.6	4	4.8
Sterile service	2.7	3	3.6	2.7	3	3.6
Utilities per year:						
Water	1.8	2	2.4	1.8	2	2.4
Sewerage	0.9	1	1.2	0.9	1	1.2
Energy	2.7	3	3.6	2.7	3	3.6
Replacement and alteration costs						
@ year 5	450	500	600	42	47	56
@ year 10	900	1000	1200	85	94	113
@ year 15	1350	1500	1800	126	140	168
@ year 20	1800	2000	2400	168	187	224
WLC risk reserves (15%) of subtotal		244		3	3.9	4
Preliminaries (12%) of subtotal		195		2.4	2.7	3.2
Decommissioning or recommissioning @ year 25 as PV		500		6.3	7	8.5
Total £/m² per year					635	

implications on running and maintenance costs, durability, functional quality, etc. The WLCC outline cost plan at this stage consists mainly of redistributing and updating the WLCC estimated in the previous phase. The WLC budget may be distributed among the major components of the project, as shown in Table 8.3.

All costs should be adjusted to account for the characteristics of the facility under design plus constructional and operational conditions and risks. All design solutions and decisions should be costed and alternatives evaluated using performance measures (quantitative and qualitative) set up by the client or wider industry; see Chapter 4. Since the majority of design decisions are taken at this stage, stakeholders have a great opportunity to influence construction and operation costs. Cost probability distributions should be produced for all major elements and future operational and maintenance costs of the project. The budget should be adjusted to all risks.

The outcome of an outline whole life cost plan is a broad cost allocation for the major components of the building, including service life, maintenance, operation, and replacement costs as shown in Table 8.3. The following assumptions might be required for the preparation of an outline WLC budget plan:

(1) *General requirements*
- reports from the previous stages
- client brief and design parameters
- outline drawings
- services requirements
- town planning issues
- project constraints.

(2) *Cost information*
- £/m^2 for all major components
- £/m^2 for services acquisition and installation
- cost report from the previous stages
- £/m^2 for annual maintenance cost for the main elements
- £/m^2 for replacement and alteration requirements
- £/m^2 for operation including energy, cleaning, etc.
- environmental costs, carbon tax
- percentage of operational costs to capital cost
- percentage of maintenance costs to capital cost.

(3) *Service life*
- service life of all major components
- failure rate and risk of failure
- component replacement frequency.

(4) *End of life*
- options for disposing of the asset
- resale value
- disposal methods
- cost reserves for end of life disposal risks.

Table 8.3 An exemplar WLCC budget estimation at outline stage.

Asset life-cycle: 25 years Discount rate: 6%	Capital cost Range estimate £/m^2			Service life	Annual maintenance and operation (M&O) as % of capital cost £/m^2			
Budget element	−10%	Base cost	+25%		As % of total capital cost	%	Cost	Annualised PV
Substructure	351	390	487	100	Fabric maintenance	6	1125	18
Superstructure	1697	1886	2357	100	Utilities	5	937	15
Internal finishes	586	651	813	10	Waste management	0.5	94	1.5
Fittings and furnishings	427	475	593	10	Cleaning	8	1500	24
Utilities	1485	1650	2062	20	Repairs and maintenance	5	937	15
External works	454	505	631	40	Life-cycle replacement	12	2250	36
Subtotal		5557			Disposal	2	375	6
Fees @ 15% of subtotal		833			WLC risk reserve	7	1313	21
Preliminaries @ 10%		556			Overheads	9	1688	27
Risk reserves @ 10%		556			Administration	3.5	656	10
Total capital cost Total annualised		7502 119					10875	173
Annualised WLCC PV £/m^2								173 + 119 = 292

8.5.4 *Whole life-cycle cost planning at scheme design stage*

WLCC planning at the scheme design stage consists of completing the brief. Decisions should be taken on what alternative is best suited to the client, based on the whole life information generated in the above stages. By this stage major decisions that influence life-cycle design, planning, construction and operation of the facility under consideration will be considered in detail. It is expected that at the scheme design stage the brief is fully developed, the architecture design is fully completed and service engineers have already coordinated their preliminary design with the architect at an early stage to test the whole life feasibility of proposed services. Approval for WLCC plan and design should be obtained at this stage. A sketch scheme should cover all the major components of the building, as shown in Table 8.4. It should be emphasised that a consistent work breakdown structure is essential to enable:

- A WLC estimate
- Capture of all costs
- Communication between project stakeholders
- Compatibility with future cost allocation and estimation.

It is important that client needs are understood so that all whole life cost elements are captured. It is expected that at this stage the WLC cost plan for the major building components will be accurately estimated and reflect client and customer needs. The following assumptions might be needed for carrying out scheme design WLCC planning:

(1) *Design service life*
   - structure
   - cladding
   - mechanical services
   - internal fabric
   - fixtures and fittings
   - external works.

(2) *Building cost*
   - approximate quantity for all elements
   - £/m^2 for retail or office space, £/bed, £/pp, etc.
   - £/m^2 for all components
   - risk allocation costs.

(3) *Operation*
   - energy
   - cleaning
   - waste management
   - utilities
   - taxes
   - staffing cost if applicable.

(4) *Maintenance and repairs*
- periodic replacement
- annual repairs
- unplanned repairs.

(5) *Disposal*
- reserve for waste disposal
- demolition methods
- site and land clean up implications
- risk reserves for end life.

(6) *Other costs*
- environmental costs
- risk costs.

An indication of how the above whole life-cycle budget is divided as a percentage between cost centres is shown in Figs 8.2 and 8.4. The figures are just used for illustration of how a whole life-cycle budget can be shared between different cost centres, but it is perfectly feasible that such information could be readily collected and developed as benchmarks for the wider industry.

8.5.5 Whole life-cycle cost planning at detailed design stage

WLC budget estimates change as design progresses due to the fact that details of element specifications and client requirements tend to be clearer as information is readily available. Thus, detailed elemental measurement and unit rates are required to estimate WLCC at this stage. Here WLCC estimation ought to be based on a detailed elemental breakdown of the project. Such a process will allow for building component selection and development of environmental specification profiles of components. The process of WLCC estimation here should merely be a redistribution of whole cost allocated previously over the range of detailed subelements. Changes may occur in the subelement total costs; however, the overall whole life cost plan should remain within the cost boundaries defined by the client at the outset of the design. All outstanding issues and decisions relating to design, specification, construction, operation, maintenance and cost should be solved at this stage. Detailed analysis and trade-offs between design alternatives and maintenance, operation and component replacement ought to be performed, especially those decisions relating to mechanical services. By the end of detailed design every aspect and component of the building must be fully designed and comply with all legal aspects including health and safety, environmental legislation and building regulation. Also all anticipated construction and operation problems must be dealt with and risk reserves put aside for all unanticipated events.

Since detailed designs and specifications are sufficiently developed, WLCC estimation ought to use approximate quantity techniques for updating the whole life cost plan derived previously. Costs for operation, maintenance

Table 8.4 WLCC budget estimation at scheme design stage.

Asset life-cycle: 25 years Discount rate: 6%	Capital cost £/m^2 (1)	Service life	Maintenance and operation costs estimate £/m^2 (2)			Periodical replacement £/m^2 (3)			WLCC PV £/m^2 (1+2+3)
Budget element			M&O as % of capital cost	M&O cost	PV of M&O	Year	Cost	PV	
Substructure	392	100	5	20	5				397
Frame	405	100	5	20	5				410
Upper floors	70	100	120	4	1				71
Roof	380	100	120	456	106				486
External walls	187	60	120	224	52				239
Windows/ext. doors	80	20	130	96	22	20	85	26	102
Internal doors	90	20	120	117	27	20	96	29	117
Internal partitions	300	10	120	360	84	20	321	100	400
Stairs	70	20	120	8	2	10 20	75 75	42 23	114 95
Floor coverings	361	10	150	541	126	20	386	120	607
Finishing	228	20	150	342	80	10 20	244 244	136 76	444 384
Ceilings	90	20	140	126	30	20	96	29	149

(Contd)

Table 8.4 WLCC budget estimation at scheme design stage (cont.).

Asset life-cycle: 25 years Discount rate: 6%	Capital cost £/m^2 (1)	Service life	Maintenance and operation costs estimate £/m^2 (2)			Periodical replacement £/m^2 (3)			WLCC PV £/m^2 (1+2+3)
Budget element			M&O as % of capital cost	M&O cost	PV of M&O	Year	Cost	PV	
Fittings	250	5	150	375	87	20	267	83	420
Furnishings	300	20	150	450	105	5 10 15 20	321 321 321 321	240 180 134 100	645 585 539 505
Mechanical services	650	20	150	975	227	20	695	216	1093
Electrical services	260	20	150	390	91	20	278	87	438
Rainwater goods	80	20	150	120	28	20	86	27	135
Public health services	219	40	150	328	76				295
Ext. works	411	40	150	616	143				554
Subtotal				5568	1297				
Risk reserves (10%)	452			547	127				579
Preliminaries (10%)	452			0					452
Total £/m^2									10255

Table 8.5 An exemplar detailed WLCC analysis for comparing between competing alternative solutions.

Study period: 25 years Discount rate: 8%		Capital cost (1)			Service life	Annual M&O @ 95 and 120 percentile (%) (2)				Periodical replacement cost @ 95 and 120 percentile							Annualised WLCC PV £/m^2 (1+2+3)		
WLC budget for dry/ wet flooring systems		Range estimate @ 95 and 120 percentile £/m^2								@ 3% annual inflation rate				Annualised cost @ 8% discount rate (3)					
		−5%	Base cost	+10%		M&O as % of (1)	£/m^2			year	£/m^2			£/m^2			Low	Base	High
							@95%	Base	@120%		@95%	Base	@120%	@95%	Base	@120%			
Wet area floor	Ceramic tile	47	50	60	50	3		1.5				0			0		2	2	2.5
	Quarry tile	38	40	48	30	5		2				0			0		2.5	2.6	3
	Rubber sheet	33	35	42	10	40		14		10 20		47 63			0.6 0.9			15 15	
	Vinyl sheet	14	15	18	15	80		12		15		20			0.3			13	
Dry area floor	Carpet	23	25	30	10	35		9		10 15		34 39			0.5 0.5			10 10	
	Laminated wood	76	80	96	20	10		8		20		144			2			11	
	Wood plank	57	60	72	30	35		21				0			0			21	
	Cork	27	30	36	10	45		14		10 20		40 54			0.5 0.7			15 16	

and replacements may be obtained from specialist suppliers, contractors, etc. if necessary. All building components should be measured and priced for procurement, maintenance, operation and disposal. Some elements may prove to have a significant cost and risk attached to them; these ought to be checked constantly through the whole life process to make sure their allocated budget stays within the margin allowed by the client. Assumptions required for the preparation of the detailed design WLC budget plan are mainly an update of the design scheme assumptions.

Table 8.5 demonstrates how a detailed whole life cost estimate may be carried out. Notice that the percentile range estimate is narrower at the end of the design stage. Economic performance indicators can also be used here to assist in the selection between alternatives. The detailed whole life cost estimate shown in this table should enable all the stakeholders to control costs due to any scope change in the project. This will assist in the assessment of risk and unforeseen events or circumstances which may arise throughout the life span of the project.

It is important to realise that such detailed whole life cost studies are expensive and it may be impractical to make such an analysis for all building elements. A detailed whole life cost analysis ought only to be used for comparison of important aspects of projects that have a large cost allocation and are deemed to potentially have large maintenance and replacement costs.

8.6 Summary

Dealing with the consequences of design decisions during the occupancy stage of a building can impact seriously on future WLCC. The opportunities to influence design decisions in the operational stage are limited and so it is vital that the risk and consequences of these decisions on the total cost of ownership of assets are identified and planned for. Therefore, it is of vital importance that an appropriate method is developed of establishing a framework at the design stage, which facilitates the synergism of WLCC, service life, environmental life-cycle assessment, and the risk associated with decisions taken.

References

Flanagan, R. & Tate, B. (1997) *Cost Control in Building Design.* Blackwell Science, Oxford.

Kirk, S.J. & Dell'Isola, A.J. (1996) *Life-cycle Costing for Design Professionals.* McGraw-Hill, New York.

9 Whole Life Risk and Risk Responses at Design Stage

9.1 Introduction

Risk analysis and management are important aspects of the WLC decision making process throughout the life-cycle of building assets. There is no such precise deterministic WLCC. In estimating the WLC budget for building assets there is uncertainty about nearly every estimate or input data that is to be used in models for computing WLC. WLC budget estimate is subject to variation due to uncertain events that may affect the process of the asset development and operation. That is why it is of paramount importance that the effect of those uncertainties on the WLC budget must be assessed along with appropriate strategies and responses for mitigating the effect of these risks. WLC risk assessment is necessary because of:

- Data used for WLCC calculation being based on estimates
- Uncertainties due to project scope
- Replacement cost and time uncertainties
- Uncertainties about operation conditions
- Uncertainties in investment parameters (e.g. rate, taxes, inflation, etc.)
- Uncertainties in project internal and external factors.

The above factors, among others, contribute to uncertainties in WLCC budget estimation at design stage. This would have a knock-on effect on the confidence of decision makers using the results of WLC for short- and long-term decisions. Uncertainties in long-term investments like building assets are significant and may lead to a total distortion of the predictions, in unknown consequences, making any decision based on WLC budget values highly suspect. Risk analysis helps in providing a level of confidence about whether the results of a WLC budget at design stage are reliable. This is particularly important when comparing competing design alternatives; if the effect of uncertainties is significantly large it will make the apparent WLCC difference between two alternatives inconclusive (Kirk & Dell'Isola 1996). Thus, one can state with confidence that all WLCC processes at design stage have some elements of uncertainty and risk attached to them. Effective identification and management of these risks is a challenging task. Hence, the purpose of this chapter is to identify whole life risks relating to design stage and strategies for responding to them.

9.2 Design whole life risk

It is universally accepted that there are four major components of risk management:

- Risk identification
- Risk analysis
- Risk responses (reducing measures)
- Risk monitoring (to ensure effective implementation of risk reduction measures).

All the above risk components are equally important and they are dynamically linked; they cannot each be addressed in isolation. Managing WLCC risk using these components will enable stakeholders to move smoothly from one component to another by identifying and understanding the possible causes of risk in the different stages of the life-cycle of assets. A successful WLC budget at design stage requires analysis of all of the risks and the establishment of clear contractual arrangements that allocate risk burdens appropriately. For this purpose the WLC design risks identification can be broadly classified into general and project-specific.

9.2.1 Design general risks

These risks are normally allocated through agreement between project stakeholders and might include:

- *Political risk*: Political risks are attributed to legislative changes covering tax regimes, laws, concession and political stability.
- *Commercial/economic risk*: Commercial risks may arise due to the fact that asset revenues will be insufficient to repay debts because the forecasted volume (e.g. number of users) will not materialise. Economic variable risks, like interest and inflation rate risks, are generally difficult to quantify with any real precision. Hence, a sensitivity analysis must be carried out to demonstrate that clients' interests are well protected against movements in the underlying interest rate and the risk that inflation increases in the future.

- *Environmental risk*: Environmental risks can be defined as the probability that an asset will have an adverse environmental impact beyond that which was permitted at the planning application stage. The consequence of this is dramatically increasing asset liabilities and total ownership cost.
- *Social risk*: This risk is concerned with the rapid changes in customer requirements due to technological advancement. Society values and norms play a vital role in social risks and in obsolescence of building assets.

9.2.2 Design project specific risks

Project specific risks are mainly associated with the phases of the life-cycle of building assets and include the following.

Risks related to the design process

The design process is highly exposed to risk uncertainty. Project design risks evolve from misunderstanding project objectives and threats to achieving these objectives. This is especially important for aspects relating to design brief, whole life objectives and parameters, design programme, WLC cost plan, time constraints and quality issues. The success of understanding and translating project whole life objectives into a reality depends on the design team's in-depth knowledge of the whole life-cycle process and associated sources of risk. This situation is further complicated by the fact that design processes are highly interactive and involve an extensive exchange of information between design stakeholders.

Risks related to costing and estimation processes

Costing and estimating risks are attributed to uncertainty in time and cost input data and may have the effect that the WLC budget allocated to the development and operation of the asset will be insufficient to build and run the asset. This is also closely related to schedule risks where the time allocated to design, procurement and construction stages is insufficient, resulting in duration overrun and associated consequences.

Risks related to contractual procedure and preconstruction decision making processes

The quality of the contractual arrangements and decisions in contract documents has an impact on how well a successful relationship develops between project stakeholders. This largely depends on the clarity and robustness of the risk allocation mechanism within the contract. Risks may not be allocated appropriately and should be transferred to those best placed to manage them. Risks due to lack of provision for dispute resolution, that may arise from contract interpretation, are quite common in the construction sector.

Risks related to construction processes

This might be looked at as the probability that the funds and time scale allocated to the project construction stage will be insufficient to complete the project. Other risks include that the construction period lasts longer and building costs are higher than expected, coupled with the usual problem that facilities are not built to the required specification. These risk aspects arise from poor WLC budget estimation, delays and failure to meet the client's aspiration and requirements. Risks of contractors, subcontractors and suppliers defaulting are also possible.

Risks related to operation processes

Operational risks have many facets ranging from risks relating to asset or component life span failure, to the asset failing to perform functionally, economically and technologically or failure of the asset to generate adequate revenue coupled with excessive consumption of resources and large budgets for unexpected repairs and maintenance.

It is also important to realise that when WLCC operational and constructional risks are transferred contractually some residual risk will remain with the client/lender. For example, if the operating costs turned out to be significantly greater than was originally forecasted it is in the best interest of the stakeholders to share this burden rather than running the risk of abandoning the project. Sensitivity analysis on capital and operational expenditure should be carried out to test if there is any residual risk. If this is the case then a fair risk allocation mechanism should be put into place and communicated clearly to all stakeholders.

Risks related to technology/obsolescence

Building assets are usually procured over a long period of time. Over this period equipment, building components and assets used in the operation of the building facility become obsolete. Also there will be a need to invest in new equipment based on new technology for the efficient operating of building assets and the delivery of associated services.

Risks related to project finance and organisation

This is the risk related to the fact that legal and managerial structures put together to develop and operate the asset will not perform as expected. The financial risk is the probability that the capital and operating costs allocated to the project are insufficient. If the project is not well planned then there is the risk of possible failure to raise funding for the project from lenders or in the market. There could also be risk relating to the fact that the asset is not generating enough revenue to repay the invested capital.

Risks related to end of life and demolition processes

This is the residual value risk and the risk of having to dispose of waste from the demolished asset. The residual risk is that which is borne by the stakeholder taking over the capital asset at the end of the service life of the facility. The significance of this is largely dependent on the type and condition of the asset under question and whether a secondary market exists for this type of asset to be functionally changed to serve other purposes. The demolition risks arise from the eco-costs associated with the process and the lack of funds allocated for this purpose.

9.3 WLC risk identification and risk response measures at design/precontract stages

WLC risk management at design stage begins with the risk identification process, which allows stakeholders to determine at an early stage the potential of internal and external risk threats to the entire asset development process. It may be argued that identification is the most important aspect of risk management for the simple reason that if a risk is not identified then there will be no reduction measures or control methods in place to deal with it should it occur.

The first step towards whole life cost risk identification is to define the asset life-cycle development environment; as we have seen in the previous chapters, the asset life-cycle environment consists of:

- Design
- Construction
- Operation and maintenance over the design life
- Future reconfigurations
- Demolition.

There are potential risks associated with WLCC at all these stages. The different types of WLCC risks at design and preconstruction stages alongside the risk reduction measures are summarised in Table 9.1 (Flanagan & Norman 1993; Edwards 1995; Chapman 2001; Jaafari 2001). The risks related to other stages are discussed in Chapters 11 and 12.

9.3.1 Risk identification

The logical process of WLCC risk management is explained in Chapters 2 and 3. Risk management for WLCC begins with the risk identification process, which allows stakeholders to determine at an early stage the potential and impact of the realisation of internal and external threats on the entire asset development and operation. It was demonstrated that WLCC risks are dynamic throughout the life-cycle of assets; hence it is necessary to identify and formulate responses to the risks that are appropriate to the life-cycle stage of the facility. The first step towards identification is to define the design project processes. It is essential to identify and define problems before putting in place measures to solve them. For risk identification purposes the design project processes consist of four stages:

- Inception/feasibility
- Scheme design
- Detailed design
- Preconstruction.

There are potential risks associated with the WLCC processes at all four stages. The different types of WLCC risk at various design stages are summarised in Table 9.1.

Table 9.1 Risk and risk responses at the design stage.

Stages	Risk identification	Possible risk response measures
Inception/feasibility	Inadequate project formulation, investigations and technical specifications	Consider making use of audit experts
	Lack of proper financial appraisal of the project	Review assumptions and use economic performance indicators
	Incomplete feasibility studies	Plan and define feasibility study tasks at very early stage
	Difficulty in capturing and specifying the client/user whole life requirements	Ensure thorough discussion with all stakeholders
	Difficulty in estimating the time and resources required to complete this stage	Consult expert and assess risk
	Difficulty in understanding and setting project objectives	Make sure that whole objectives are quantified and explained to clients
	Whole life parameters not identified at outset	Make sure that WLC parameters are part of feasibility studies
	Ill-defined whole life financial limits	Consider using financial experts
	Responsibilities between stakeholders ill defined	All participants should be part of the decision process with unity of goals
	Team composition is inadequate and unbalanced	Make sure that expert skills are identified at early stage
	Ill-defined project quality parameters	Make sure that quality parameters are included in the feasibility studies
	Ill-defined whole life assumptions	Consider checking WLC assumptions with expert before using them
	Inadequate review and preparation of planning and environmental application	Prepare and submit all necessary documents and feasibility study report in a timely manner to government departments for approval

(Contd)

Table 9.1 Risk and risk responses at the design stage.

Stages	Risk identification	Possible risk response measures
Outline/sketch design	Ill-defined design processes and inadequate programming	Undertake preproject planning
	Inexperienced design team	Consider subcontracting part of the work or providing training
	Change of project whole life objectives	Make sure that WLC objects are clearly defined and validated by experts
	Lack of communication and withholding of information between stakeholders	Maintain good relationship and continuous communication with stakeholders
	Planning and environmental constraints	Discuss and report in a timely manner to government departments for approval
	Inadequate site information and analysis	Preplan site data collection and validate analysis results
	Inadequate definition of whole life-cycle parameters	Validate WLC assumptions with experts before use in the analysis
	Misinterpretation of user/client requirements/expectations	Check with stakeholders that your interpretation matches their expectations
	Unclear or ill-defined roles and responsibilities	Stakeholders should be part of the decision process with unity of goals
	Whole life cost plan inadequately prepared	Undertake WLC preplanning and introduce control mechanisms
	Whole life budget based on false assumptions	Introduce checking mechanisms and validate assumptions with experts
	Operational strategies not considered	Make sure that operational strategies are part of WLC planning and objectives
	Unproven whole life design solutions adopted	Ensure thorough scrutiny and checking of adopted solution
	Project brief unclear	Carry out brief audit, discuss and resolve all unclear issues with stakeholders
	Allocated resources are inadequate	Check planned resources with experts and provide adequate contingency funds
	Incomplete design scope	Ensure thorough scrutiny of draft scope of work

(Contd)

Table 9.1 Risk and risk responses at the design stage.

Stages	Risk identification	Possible risk response measures
Detailed design	Large choices of design solutions of varying whole life parameters	Consider only solutions that are relevant to WLC objectives
	Compliance with standards and codes of practice	Make sure that all relevant regulations are identified and complied with
	Difficulty of integrating and exchanging whole life design information	A task force may be constituted for better coordination
	Lack of design coordination	Consider using an independent project design manager
	Lack of adherence to whole life budget	Put in place control measures
	Late completion of design	Ensure that time control mechanisms are in place and adhered to
	Design is too complex and novel	Allow for time and cost contingencies
	Environmental and planning conditions imposed on design	Discuss issues with relevant authority and allow for additional time and cost
	Change of whole life parameters	Make sure that WLC assumptions are defined and validated at the outset
	Design errors	Undertake predesign planning to minimise design errors
	Inadequate coordination between designers, structural and M&E consultants	A task force may be constituted for better coordination
	Incomplete whole life-cycle cost plan	Make sure that WLC plan is validated by experts
	Errors and omissions in whole life assumptions	Validate assumptions with a panel of experts before use
	Late planning application	Plan, prepare and submit in a timely manner

(Contd)

Table 9.1 Risk and risk responses at the design stage.

Stages	Risk identification	Possible risk response measures
Detailed design	Planning and environmental approval conditions not complied with	Consider validating all design with expert before approval
	Change in whole life cost parameters due to external environment (inflation, taxes, etc.)	Make sure that there are adequate contingency reserves to cover such eventuality
	Design does not reflect life-cycle conditions	Use control mechanism to incorporate WLC objectives in design
	Production information not to agreed quality standard	Consider using quality assurance systems
	Ineffective whole life cost control parameters	Make sure that WLC assumptions are defined and validated by experts
	Delay in design approval	Make sure that control mechanisms are in place
	Inadequate design project management	Consider using a task force for better management or providing training
	Lack and delay of data and information from manufacturers and suppliers	A task force may be constituted for better coordination
	Complexity of design	Consider using a task force of experts and allow for time and cost contingency
	Long waiting time for approval of drawings	Make sure that drawings are submitted very early
	Design changes by owner or his agent	Make sure that owner objectives are clearly identified and understood by all
	Errors, omissions and discrepancies in design	Consider using quality assurance and total quality management systems

(Contd)

Table 9.1 Risk and risk responses at the design stage.

Stages	Risk identification	Possible risk response measures
Preconstruction	Design documentation not suitable for selected procurement type	Organise for appraisal/vetting of design documentation by experts
	Delay in design approvals	Prepare and submit all necessary design information on time for approval
	Errors in design information	Consider design liability insurance
	Change in whole life economic parameters	Allow risk contingencies
	Inadequate procurement planning	Use expert to undertake preprocurement planning to avoid disputes
	Contract documents are inadequately developed	Consider using experts for vetting all contract documents
	Whole life-cycle budget is not clearly defined	Make sure that WLC budget is defined and validated by experts
	Unclear or insufficient procurement schedule	Undertake preprocurement planning to maximise chances of success
	Escalation of tendering costs	Make sure that cost control mechanisms are in place and make allowances
	Unclear risk allocation mechanisms	Have clear contractual terms and conditions for risk allocation
	Unbalanced whole life risk allocation in the contract document	Devise unambiguous and agreed WLCC risk sharing code at the time of contract

(Contd)

Table 9.1 Risk and risk responses at the design stage.

Stages	Risk identification	Possible risk response measures
Preconstruction	Unknown capability of bidder	Use only approved list of bidders
	Unclear guidelines on change procedures	Seek incorporation of terms for change procedures in the contract clauses
	Lack of provision in the contract for dispute resolution	Provide comprehensive conflict resolution clauses in contract
	Lack of training and experience to manage contract during procurement stage	Hire competent project management team or provide training
	Lack of understanding of stakeholders' businesses	Ensure openness of communication
	Lack of common objectives and vision between stakeholders	Provide clear definition of each stakeholder's scope of work or change the team
	Adversarial relationship	Maintain good relationships with all project stakeholders
	Poor management of procurement costs	Adopt proper quality cost control procedures and supervision
	Inadequate staffing of the procurement process	Hire competent teams
	Unsuitable form of contract/partnership	Consider using experts for selecting an appropriate form of contract
	Unclear whole life-cycle project deliverable; cost, service life and quality of services	Provide a contract for the definition of each WLCC objective

9.3.2 Inception/feasibility risks

Identifying risks associated with WLCC at the inception/feasibility stage involves the definition of the risk, frequency and consequences. Inception/feasibility risks arise mainly from poor whole life project objectives definition. Such risks may arise from both internal and external sources. External threats may involve client business uncertainty, market volatility and political and legal aspects of the proposed asset. At this stage these factors may have little effect on the design project, because stakeholders' financial commitment to the project is not large, hence the loss from such risks can easily be recuperated.

Internal threats to project design at this stage may come from poor definition of project design parameters. This includes all aspects relating to quality, financial limits, poor project formulation, design team experience and interaction, as shown in Table 9.1. These threats can damage the prospect of a good return on the investment. It may also lead to a project design based on wrong assumptions and user requirements.

9.3.3 Outline/scheme design risks

At this stage, the focus of risk identification is on the process of consolidating and translating the project brief and whole life-cycle objectives into a physical design. The way these two key issues are interpreted will have a significant impact not only on the design processes, and quality, but on the whole asset life and beyond to stakeholders' business processes and user satisfaction with the asset. Thus, it is fundamental to examine the brief, whole life objectives, assumptions, programme, WLC budget and all other issues relating to quality and operation of the designed asset. Other risks at this stage arise from the characteristics of the design process and activities. The design activities in the RIBA plan of work can be used as a guide for risk identification. Table 9.1 lists the sources of risk that may have an impact on project outcome at this stage. The most significant risks here can arise from the lack of balance and experience of the design team, unclear definition of roles and responsibilities, planning approval issues, approaches to design, information availability and interpretation, unclear definition of project brief and unclear whole life cost assumptions.

9.3.4 Detailed design

Here, the focus is on the design risks in obtaining a final decision on every aspect related to design, specification, construction, operation and WLCC issues. The challenge here is in the integration of the project design teams who are participating in the project development. Risks here mainly arise from lack of effective communication and unclear responsibilities between design stakeholders. Further risks could be attributed to the sequencing and exchange of information, the impact of external agencies (e.g. local authority) and the management of changes to the brief and WLC objectives and assumptions.

This has the consequence of generating other design risks like design time and cost overruns, delay in making key approval decisions, errors in design, etc. It is also worth stressing here that technology selection risks are very prominent at this stage. These risks arise from the enormous choice of materials, the technology of varying whole life attributes, service life, durability, environmental impacts, and aesthetic characteristics. A detailed list of the sources of risk at this stage is shown in Table 9.1.

9.4 WLCC risk categorisation

There are various ways to categorise and estimate the effect of risk factors that affect WLCC at design stage. Provided in Table 9.1 are the risk factors associated with projects at design stage. They are categorised according to design stages. The categorisation method and risk factors may change depending on risk analyst preferences and the type of project involved. For ease of analysis the risks shown in Table 9.1 are further clustered into:

- WLC estimate risk
- Technical risk
- Financial and economic risks
- Market risk
- Organisational risk
- Operation risks
- Schedule risk
- Political risk.

9.5 Design WLCC risk quantification

Whole life-cycle costing risk can be looked at as the probability that the WLCC estimate is sound and accurate. WLCC risk may be attributed to uncertainty resulting from the use of data and budget estimation methods. Risks and mechanisms to deal with this vary according to the asset's life-cycle. WLC information changes with time. Even when the status of investment or asset objectives is well defined they could change over time. This is because investors' business objectives change over the life-cycle of an asset due to the external business environment. This could lead to previously unknown WLC exposure risks. Risks that are perceived to be insignificant at the outset of an asset development may pose new threats. Against this complexity and uncertainty surrounding the development and operation of building assets, the challenge is to pursue WLCC risk estimation earnestly and to look for opportunities to reduce the consequences of the risk and improve the asset's value. Thus, WLC budget conceptualisation, planning and implementation are a complex, dynamic and evolving process. If we fail to recognise this fact then many aspects of risk will be ignored and the investment potential will not be realised. That is why WLC risk should be managed on the basis of asset/business strategic objectives, which are subject to change on a flexible basis over the entire life-cycle of the asset using a continuous real time risk and uncertainty reduction measures within an integrated WLC framework, as shown in Fig. 8.1.

Performing WLC risk analysis can be viewed as a mechanism to create and adjust a WLC budget estimate for the probability of risk occurrence. Risk here addresses the probability of WLC estimate variation occurring and the consequences surrounding that. WLC uncertainty is the confidence we have in the generated WLC computation. By the very nature of WLC calculation into the future, there exists an amount of risk and uncertainty with all WLC decisions. The actual WLC budget outcome might not be predictable with certainty. But the probabilities of the varying WLC budget may be known from simulation or other methods. Analysts should be able to determine the risk from these probabilities. As long as the risk is identified, it can be managed and controlled.

Once risks are categorised or clustered they must be evaluated. The purpose of evaluation is to realise a quantification of risk. Risk evaluation should be carried out according to two different approaches: by the evaluation of resulting consequences (deterministic approach) or the evaluation of risk event occurrence probability (probabilistic approach). It is necessary to rank the identified risk factors. The ranking process will help decision makers in concentrating their efforts on the factors that are deemed to have high risk impacts. Risk responses will then be developed to deal with the anticipated consequences.

The methods used to evaluate risk can be classified into two broad categories: qualitative and quantitative.

9.5.1 *Qualitative risk evaluation*

Once risks are defined and identified, a high, medium and low scale might be used to assess the effect of each risk factor on a building asset objective, in our case a WLC budget. Then, after the analysis, each factor's numerical score can be converted to a cost variation of WLC budget element based on the corresponding scale shown in Table 9.2. Alternatively a simple total score for each factor might be used as illustrated in Table 9.3. Finally, a risk ranking

Table 9.2 Risk effect scale

Score	Risk level
1	Low
2	Medium
3	High

Table 9.3 Risk outcome scale

Weighted average score	Risk impact
1.0–1.5	Low
1.6–2.5	Medium
2.6–3.0	High

Table 9.4 An example of qualitative risk assessment.

Risk type	Risk weight %	WLC budget	Design time	Design cost	Operational cost	Construction cost	Design process
WLC estimate risk	30	2.5	2	1	3	2	1
Technical risk	10	1	3	3	2	3	2
Financial and economic risks	20	1.5	1	2	3	2	1
Market risk	5	1	1	2	3	2	1
Organisational risk	5	1	2	2	2	2	3
Operation risks	10	2	3	1	2	1	1
Schedule risk	10	1.5	2	3	2	1	1
Legal risk	10	1	2	1	1	2	1
Weighted average risk	100	1.5	1.95	1.7	2.5	1.9	1.2
Risk impact level		Low	Medium	Medium	High	Medium	Low

summary is developed to assess design WLC budget risk graphically and numerically, as shown in Table 9.4. It is also possible to use data in Table 9.4 to carry out a sensitivity analysis and Monte Carlo simulation for allocating whole life risk contingency reserves.

In carrying out qualitative WLC risk evaluation, analysts may consider the following inter-related issues (some are modified from Smith 1999):

- Definition and description of the risk
- The whole life stages of the design when it may occur
- The aspects of the design and WLCC that could be affected

- The influencing factors that initiated the risk
- The relationship with other factors
- The likelihood of it occurring
- The severity of it and how it could affect the design
- What are the possible response strategies?

9.5.2 Quantitative methods

There are a variety of quantitative risk assessment methods. WLC design risk can be evaluated by applying a wide range of quantitative techniques. Quantitative WLC design risk assessment might be defined as the application of any technique that deals with probability in some mathematically sound basis for assessing WLC risks. Some of the widely used quantitative risk assessment methods that may be useful in evaluating WLC design risks are:

- Bayesian analysis
- Monte Carlo simulation
- Delphi methods
- Utility theory
- Analytical hierarchy process method
- Portfolio theory
- Value @ risk theory.

The aim here is not to discuss these techniques in detail as Chapter 5 explains some of these methods in depth. This section concentrates primarily on range estimate of the effect of risk factors on WLC at design stage. One way to get a better understanding of the effect of risk factors of WLC is to use the sensitivity analysis technique to compute the range of possible outcomes. We can estimate the upper and lower bounds of the effect of a risk factor by recalculating the effect with the lowest and highest likely WLC estimate. The process starts with identifying the risk influencing factors. As explained above, the risk factors should represent major WLC related issues at the design stage, and they should be independent of each other.

Typical risk factors are shown in Table 9.1 and categorised in Section 9.4 into eight main groups. The influence of these factors could be first established qualitatively based on the criteria defined in the qualitative risk assessment section. Then the effect of the influencing factors on WLC issues is determined. The results of this should be a WLC-influence matrix for the WLCC base estimate. A typical WLCC-influence matrix is shown in Table 9.5. The net effect of all of the estimated ranges that are due to each risk factor is computed at the bottom of the WLCC-influence matrix. Also it is possible to compute the effect of risk factors on each WLCC component row-wise across the WLCC-influencing matrix, as shown in Table 9.5. The advantage of assessing risk effects in this way is that it assists in evaluating risk for each WLCC component. However, this approach may complicate the risk estimation approach if a large number of WLCC components are analysed. By computing the effect of risk factors on WLCC column-wise instead of across the matrix rows, a risk

Table 9.5 Risk influence matrix.

Risk type	%	Financing cost	WLC budget	Design time	Design cost	Operational cost	Construction cost
		V1	V2	V3	V4	V5	V6
WLC estimate risk	+		5			10	5
	−		10			20	10
Technical risk	+				6	10	5
	−				15	15	10
Financial and economic risks	+	5	5				
	−	10	10				
Market risk	+	3					
	−	5					
Organisational risk	+			4	5		
	−			10	6		
Operation risks	+						
	−						
Schedule risk	+			5	5		
	−			8	10		
Legal risk	+		2	3		3	5
	−		5	5		6	4
Total risk impact	+	8	12	12	16	23	15
	−	15	25	23	31	41	24
Risk adjusted base estimate							

estimate can be computed for the WLC estimate as a whole. Further analysis is then carried out only on the total WLC estimate and associated risk-influencing factors.

The estimation of the variation caused by the risk-influencing factors is based on the following assumptions:

(1) A range around WLCC base estimate is a measure of the variation of the WLCC
(2) A continuous probability distribution is used to model WLCC variation due to risk

(3) Upper and lower limits of the variation probability distribution are defined by the range estimate
(4) The most likely effect is no change to the WLCC base estimate (i.e. influencing risk factors have no significant impact on WLCC)
(5) Risk factors are considered independent. Hence the summation of the effects.

Instead of using the method shown in Table 9.5, some analysts may prefer to use other mathematical functions to model the variation of WLCC due to risk-influencing factors. Once the input risk variations are modelled, Monte Carlo simulation can be used to determine the effect of risks on WLCC estimates at the design stage. The purpose of using it here is to assess the effect of risk factors on the base WLC estimate using the data from the completed WLCC-influencing matrix. The relationship between WLC and risk variations shown in Table 9.5 can be modelled using the following equation:

$$\text{Risk adjusted WLCC base estimate} = \text{WLCC base estimate} \pm \text{risk impacts}$$

Application of Monte Carlo techniques to the above equation will produce a range of WLCC forecasts. The simulation will cause all reasonable combinations of the risk variable coefficients to be combined in the above equation. Monte Carlo simulation randomly combines the effect of the influencing factors with the base WLCC to produce a risk-weighted WLCC that is described by input range or variation due to risk factors. Therefore, the deterministic base WLCC is contained within the range of the risk-weighted WLCC. This will yield an output curve that represents all probable WLCC outcomes based on anticipated risks.

9.6 Design risk response measures

Once risk has been identified and its impact significance measured, the analyst has to determine the most effective manner of handling the risk concerned.

9.6.1 *Risk avoidance measures*

Risk avoidance refers to elimination of either risk sources or their consequences. This can be achieved through:

- Selecting different design solutions
- Taking less risky decisions
- Changing construction technology
- Using different procurement methods
- Change to client/user requirement.

9.6.2 *Risk reduction measures*

Risk reduction measures refer to two aspects. The first is related to the reduction of probability risk occurrence. The second is that the risk might occur but measures for impact reduction are put in place to deal with the resulting consequences. WLCC risks can be reduced by (some are modified from Smith 1999):

- Obtaining additional WLCC information
- Performing additional analysis
- Allocating additional resources
- Improving whole life-cycle communication and managing life-cycle phases and stakeholder interfaces
- Using expert consultants to reduce risks
- Using formal planning and control procedures
- Educating stakeholders and project participants
- All project stakeholders should be part of the WLCC decision processes with unity of vision and objectives
- Using real-time risk evaluation and management processes
- Forming whole life strategic partnerships
- Agreeing whole life-cycle objectives and the precise meaning of terms such as service life, WLCC, cost, price, value for money and risk allocation
- Agreeing terms for change
- Defining clear WLC responsibilities
- Agreeing at very early stages performance and quality indicators
- Establishing common ground and single WLC focus
- Agreeing dispute resolution procedures at very early stages of asset development.

9.6.3 *Risk retention measures*

The level of risk and uncertainty assumed will vary according to the project design. If risk is retained, the risk analyst has to demonstrate that sources of finance have been secured with respect to the various types and sizes of loss and that any risk retained is within the capability of the project or stakeholders' finances to absorb. Retaining the cost of the risk internally by stakeholders is more economic than transferring it to an insurer, but this may have to be balanced against the additional costs that may arise from claims and administration of the risk. Some analysts suggest that risks that are acceptable for retention are only those that occur frequently and have little impact or small loss. However, this is not the case in PFI and PPP projects where the risk allocation is based on the principle that risk is better placed with those that are able to manage and absorb the impact of the consequences. The cost of risk retention may include:

- *Retained losses and payment of damages with respect to liability exposure*: Other losses are due to business interruption costs, loss of revenue, loss of market share, loss of production, and cost of risk management
- *Insurance premiums*: This is related to the increase of insurance premiums and insurance indemnity
- *Cost of risk control*: This may include the cost of financing different packages and expenditure related to procedure put in place to control and manage retained risk.

9.6.4 Risk transfer measures

The extent to which risk is transferred is often dependent upon the nature of the exposure concerned. The principal rule of thumb in deciding if a risk should be transferred is whether the receiving stakeholders have the managerial and financial capacity to manage, control and withstand the impact of risks. Risk transfers can be looked at from three aspects. Firstly, whole life-cycle processes or activities that have a high risk can be transferred to those parties who have expertise to control them. Alternatively, they may be retained but the financial risk is transferred through insurance and other suitable mechanisms. The third aspect is a combination of insurance transfer and retention. Risk transfer or allocation is an area where there are often dispute problems between project stakeholders. This is why appropriate mechanisms for whole life-cycle risk transfer should be put in place. The following considerations should be taken into account when risk is transferred:

- Contract allocates risk appropriately
- Contract terms are enforceable
- Optimum transfer of risk, such as the transferee is able to manage the risk
- Premium for accepting risks
- Scope and extent of the risks being transferred.

It is important to stress that not all WLCC risks may be suitable for transfer. It is imperative that stakeholders continually monitor risks to which they may be exposed, and take appropriate steps to manage them.

9.7 Summary

WLCC models that incorporate well-developed risk and risk identification frameworks have distinct advantages over conventional deterministic models when making decisions throughout the whole life-cycle of building assets. It could be argued that there is no such thing as a precise deterministic WLCC as it fails to provide knowledge and responses to design risk. Uncertainty exists throughout the WLCC budget forecast, which is subject to the variation that may affect the process of the asset development and operation. It is therefore of paramount importance that the effect of those uncertainties on the WLCC budget forecast is fully assessed and quantified along with appropriate strategies and responses for mitigating the effect of these risks. In the next chapter we shall put theory into practice and look at how these techniques can be applied in the design and selection of mechanical and electrical services.

References

Chapman, J. (2001) The controlling influences on effective identification and assessment for construction design management. *International Project Management*, **19**, 147–60.

Edwards, L. (1995) *Practical Risk Management in the Construction Industry*. Thomas Telford, London.

Flanagan, R. & Norman, G. (1993) *Risk Management and Construction*. E. & F.N. Spon, London.

Jaafari, A. (2001) Management of risks, uncertainties and opportunities on projects: time for a fundamental shift. *International Project Management*, **19**, 89–101.

Kirk, S.J. & Dell'Isola, A.J. (1996) *Life-cycle Costing for Design Professionals*. McGraw-Hill, New York.

Smith, N. (1999) *Managing Risks in Construction Projects*. Blackwell Science, Oxford.

10 Whole Life-cycle Costing of Mechanical and Electrical Services: a Case Study

10.1 Introduction

Mechanical and electrical (M&E) services are one of the fastest developing forms of technology within building, with increasing investment of over 15% in the last 10–15 years (Cassidy 2000). The 1980s saw unprecedented growth in the M&E sector. While the increase was parallel with private office building, which almost doubled in value between 1987 and 1990, the value of M&E also increased as a proportion of total building cost, led by higher levels of specification. Despite this trend, the downturn in construction output between 1990 and 1994 was dramatic (Burgoyne 1996), with workload reduction for M&E up to 40%. However, recently the market has shown an upward trend since the late 1990s. Continued growth in the M&E sector is forecast, at least around 2% annual increase in real terms for the next four years (Building Services 2002). The cost for mechanical services can fall between 10% and 50% of the total construction cost, depending upon the desired function, sophistication, complexity and prestige of the building. It is not just the initial capital cost that a designer has to take into account when choosing the mechanical services systems. The operational costs, maintenance costs and life expectancy have to be considered. Mechanical services are usually the most expensive element of any building. They involve not only major capital investment but also significant expenditure on maintenance, renewal and operation. A relationship between building function and M&E costs was identified by Brown (1987). The study also found that the allocation of the total services cost among the various services elements could not be predicted accurately from knowing the building's function. Therefore, only totals for the complete M&E services could be predicted as a percentage of the total building cost.

Building form is another issue that has been investigated to see if it has a significant bearing on the cost of M&E services. Brandon (1978) identified a number of descriptors of building form that may be useful for determining M&E services cost. These include plan shape, number of storeys, boundary coefficient, average storey height, percentage of glazed area, and plan compactness. However, Swaffield and Pasquire (1999) found that other forms of descriptors could be more useful for M&E services cost estimates.

Horizontal distribution volume and internal cube were found to be the most useful variables.

It is not just the initial capital cost that clients have to take into account, but also operational and maintenance costs and life expectancy. Operational and maintenance costs play an important role in the procurement of mechanical services. M&E repair and maintenance in the private industrial and commercial sectors has grown dramatically since the late 1990s. Bull (1993) stated that 60% of the life-cycle budget is spent on repairs and maintenance. Hence, it is not only the initial capital costs of installation for the M&E systems that should be compared, as some systems may be cheaper to acquire and install but are more expensive to operate. It is necessary to differentiate between best value and lowest initial price.

Environmental issues also need to be checked, such as equipment compliance with current and future legislation in terms of CO_2 emissions, ozone depletion and health and safety. These will have a bearing on the initial capital cost and the whole life costs of various systems. Most of the existing methods for forecasting these costs are based on deterministic methods ignoring the inherent uncertainty and variability of the real world, for which a probabilistic methodology is better suited. For this purpose, simulation is widely accepted and recognised by practitioners and researchers as a modelling tool for dealing with uncertainty. It is also widely acknowledged that the quality of a simulation model's results is strongly correlated to the quality of the input probability distribution functions. Decisions made during the design can have a significant impact on the future operating and maintenance costs of buildings. Hence, in this case study an attempt is made to present a new concept for modelling and analysing the economic viability of air-conditioning (AC) systems using risk methods. It is based on assessing the characteristics of the probability distribution of the unit cost per m^2 for acquiring and operating four different AC systems. The purpose of the analysis is to obtain the parameters of the theoretical probability distribution functions that best model the cost of installing and running four different air-conditioning systems over 20 years. The Anderson goodness-of-fit tests are used for selecting and validating generated distributions. The characteristics of generated distributions are presented and discussed.

10.2 Modelling the whole life cost of air-conditioning systems

From a purely financial perspective total costs for acquiring AC systems can be defined under four different cost centres. These are: initial capital costs; major asset replacement costs; subasset replacement costs; and planned preventative maintenance costs and energy costs (Cassidy 2000). The model developed here is based on the assumption that these cost centres and the expectancy life of the AC components are randomly distributed according to one of the theoretical distribution forms. This requires that each cost centre element be treated stochastically and the cost of operating AC systems is represented as probability distribution function. The total cost of operating AC systems is considered to be variable from one year to another depending on inflation and interest rates. Hence, if the PDF or cash flow profile of each AC system acquisition and running is known or can be simulated, the total PV of acquiring and operating AC systems can be estimated using the following equations:

$$PV = CAP_c + \sum_{i=1}^{n} (AR_c + SR_c + EG_c + PM_c)/(1 + r)^i \quad (1)$$

Where:
CAP_c = the initial acquisition costs
i = time or period of study
AR_c = major asset replacement costs
SR_c = subasset replacement costs
PM_c = planned preventative maintenance costs
EG_c = energy costs relating to the running of AC systems.

The second term of equation (1) can also be expressed in the following formula:

$$AC_{vr|q} = \sum_{i=1}^{n} (AR_{c|q} + SR_{c|q} + EG_{c|q} + PM_{c|q}) \times P\{q\}/(1 + r)^i \quad (2)$$

Where $q = 1 \ldots\ldots, m$ represents possible events (costs), AR_c, SR_c, EG_c and PM_c are running costs respectively in period i due to the occurrence of q, $P(q)$ is yearly probability of occurrence of a particular event (in this case the probability that q occurs), AC_{vr} is the expected cost in period i and i is the time or period of analysis. AC_{vr} is the variable cost of running the acquired AC system. Equation (2) expresses the yearly stochastic nature of the AC acquisition and operation. This equation is useful for carrying out a sensitivity analysis to find out the parameters that influence the viability of AC systems. As described above, AR_c, SR_c, EG_c, PM_c and r are random quantities, and hence AC_{vr} is a function of several random parameters. The composition of AR_c, SR_c, EG_c and PM_c may have a significant influence on the form of the PDF of yearly and total PV computation.

Initial capital costs include the costs for equipment, installation commissioning, design fees and legal fees. With regard to energy costs, energy efficiency of AC systems is becoming increasingly important as it can account for the largest single element of the energy requirement of a building. In the near future, energy tax might be introduced. It may be more cost effective to install a system with higher initial cost but which provides greater energy efficiency. Operational and maintenance costs include cleaning, servicing and repair.

Maintenance cost is often overlooked in any calculation and can be very costly dependent on the complexity of the system and the AC system. Flexibility of the installed AC systems should also be assessed in relation to the occupier or user, the operator and/or maintenance personnel. Many systems have standard controls, which are simple in concept yet sophisticated in nature and can combine user friendliness with full technical diagnostics. Any planned replacement of plant or other components at the end of useful life should also be considered. In this case study replacement costs are subdivided into major asset replacement (items of plant and equipment that have a life span over 10 years) and subasset replacement (this includes the periodic replacement of AC components, like motor drivers, etc.).

10.3 Data and methodology

To study the economic viability of AC systems, PV financial indicators are used as effective measures for selecting a viable investment in AC systems. To compute PV values, accurate information regarding the input determinants in equations (1) and (2) is required. The state of knowledge of AC systems over their life expectancy does not provide sufficient data or knowledge to assist financial analysts in estimating deterministically the total cost of acquiring and running AC systems. Because of this lack of information Monte Carlo simulation is used here to evaluate the economic viability of AC systems. Monte Carlo simulation allows the use of probability distribution for each of the variables in equations (1) and (2), enabling the simulation model to account for uncertainty and variability in the cost determinant factors of AC systems. Monte Carlo simulation also allows the input variables that are not independent of each other to be modelled and incorporated into equations (1) and (2). Since there is a lack of data, there are two probability distributions that are suitable for modelling input variables in the absence of sufficient information or when the system to be modelled does not exist. These two probability distributions are the triangular and the beta distributions (Law & Kelton 1991; Back *et al.* 2000). In this case study triangular distribution is used to develop assumptions about the minimum, maximum and modal values of the random input variables. Assume that x represents the annual planned maintenance as a random quantity with triangular distribution. Then the probability density function of the yearly maintenance cost of each AC system can be modelled using the following mathematical expressions:

$$f(x) = \begin{cases} \dfrac{2(x-a)}{(b-a)(c-a)}, & a \leq x \leq b \\ \dfrac{2(c-x)}{(c-b)(c-a)}, & b < x \leq c \\ 0, & \text{elsewhere} \end{cases} \tag{3}$$

Where $a < b < c$ are the parameters of the distribution as shown in Fig. 10.1. The expected PV value of each AC system is computed using the following formula:

$$E(x) = \frac{a + b + c}{3} \tag{4}$$

According to the engineering statistics handbook the triangular distribution leads to a less conservative estimate of uncertainty, i.e. it gives a smaller standard deviation than the uniform distributions. Hence, the estimate of the lower and upper limits can be based around the expected value *E(x)*. If it can be assumed that *E(x)* is the true average of the cost of each AC system, then the next assumption that needs to be made is how close the upper and lower limits are to the mean or *E(x)*. In any case, whatever the form of the distribution that represents the input cost of each AC system, an estimate for the upper and lower limits can be obtained through the mean–variance method. In this case study the mean of the distribution function of the cost for each AC system input variables is assumed to be the estimated value of the base case, i.e. *E(x)*. The coefficient of variance of the lower limit for each AC system input variables is assumed to be 0.1, so that the standard deviation will be:

$$\sigma_1 = 0.1E(x) \tag{5}$$

Hence the lower limit value for each AC system input variables is computed using the following formula:

$$E(1) = E(x) - 0.1E(x) \tag{6}$$

whereas the coefficient of variance of the upper limit for each AC system input variables is assumed to be 0.2, so that the standard deviation will be:

$$\sigma_p = 0.2E(x) \tag{7}$$

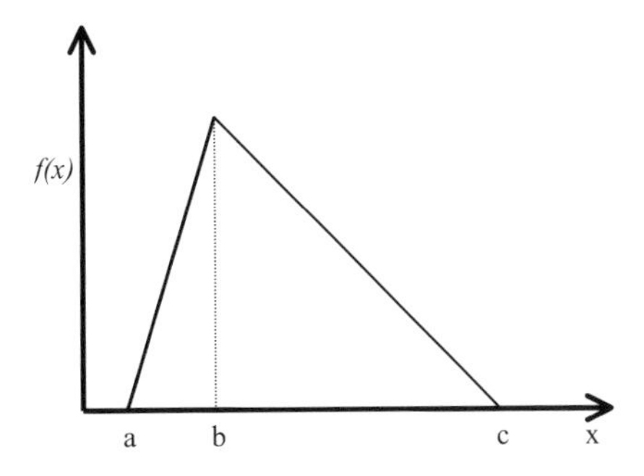

Fig. 10.1 Triangular distribution function.

Hence the upper limit value for each AC system input variables is computed using the following formula:

$$E(p) = E(x) + 0.2E(x) \tag{8}$$

The above formulae suggest that if an input variable has a high uncertainty the coefficient of variance of this variable can increase depending on the perceived uncertainty associated with the variable.

The most likely values in Table 10.1 were extracted from actual cases. These values along with the assumption made in equations (5) to (8) were used to estimate the parameters of the probability distribution function of the input variables. The capital cost of the four AC systems is based on the actual data reported by Cassidy (2000). The cost of running the four AC systems is based on assumptions reported in the same publication. These values can be replaced by more accurate data if and when better information is available to cost analysts. The model used to represent input data follows the three-point estimating procedure (triangular distribution). This method requires the analyser to set a lower and an upper limit to a most likely base estimate for every AC system under investigation, the item should be costed according to the most optimistic and pessimistic conditions, respectively. The degree of variability is indicative of the level of certainty about the cost under investigation. Monte Carlo simulation is applied to determine the distributions of PV, given that the probability distribution of each variable is known. The probability distributions of the AC determinant parameter can be developed from experience of similar occupancy costs in approximately similar conditions. The triangular distribution was selected for modelling the occupancy cost input variables. The rationale behind the selection of triangular distribution is discussed in Fente *et al.* (2000), Law & Kelton (1991), Wall (1997) and Chau (1995). The parameters of the triangular probability–density function can be estimated using expert subjective judgement (Chau 1995) or from historical data using moment matching, maximum likelihood estimation, and least-squares fit of the cumulative distribution function (AbouRizk *et al.* 1994).

In this case study equations (3–6) are used to define the triangular distribution parameters. If an input variable has a high uncertainty the coefficient of variance of this variable can increase depending on the perceived uncertainty associated with the variable. The most likely values in Table 10.1 are extracted from actual cases. These values, along with the assumptions made about the upper and lower limits of each variable in equation (1), are used to estimate the parameters of the probability distribution function of the input variables. Goodness-of-fit tests are used to select the best probability distribution function that represents PV output variables. Monte Carlo simulation is used to generate the distribution of possible PVs. It takes samples from the input variable distributions and evaluates the corresponding NPV that is a function of these variables. The process is repeated for 2000 iterations and the resulting PV1, PV2,..., PVn are used to obtain the cumulative distribution of PV. Obtaining PV values in this way is subject to estimation error resulting from

Table 10.1 Input parameters.

AC systems Variable air volume with perimeter heating (VAV) Variable refrigerant flow (VRF) Fan-coil units (FCU) Chilled ceilings (CC)	AC installation costs (£/m²)	Electrical work costs (£/m²)	Associated costs (£/m²)	Every 15 years major asset replacement (£/m²)			Every 5 years subasset replacement (£/m²)			Every 5 years subasset replacement (£/m²)			Annual energy costs (£/m²)		
				0.1E(x)	E(x)	1.2E(x)	0.1E(x)	E(x)	1.2E(x)	0.1E(x)	E(x)	1.2E(x)	0.1E(x)	E(x)	1.2E(x)
VAV	228	7.5	96	117	130	156	39	42	50	56	62	74	197	219	263
VRF	166	6	66	191	212	254	50	55	66	108	123	147	122	136	163
FCU	183	7	72	80	89	107	56	62	74	98	109	131	123	137	164
CC	229	3	88	43	48	58	43	48	58	27	24	29	56	62	74

sampling error, and inappropriate discount and inflation rates. Therefore, the reliability of the generated distribution has to be assessed. In this case study, it was done with the Anderson goodness-of-fit test, which checks if the observed data could have originated from a theoretical distribution with the estimated parameters.

10.4 Results and discussion

Table 10.2 shows the results of matching the forecasted PV of each AC system to a theoretical distribution function. The matching is based on the Anderson goodness-of-fit test. A total of 32 continuous distributions were assessed. Those which best represent the forecast data along with their characteristics are shown in Table 10.2. Table 10. 2 shows that the gamma distribution is ranked first for the VAV AC system. But the random walk distribution was found to be the best fit for the VRF AC system. It was found that the fan coil unit and chilled ceiling AC system is best represented by the Johnson SB distribution.

Figure 10.2 illustrates the preliminary analysis carried out to check for the goodness-of-fit of the selected distribution. Figure 10.3 shows distribution function differences plotted for the forecasted PV values for the investigated AC systems. The graphs show that the vertical differences between the sample distribution and the theoretical distribution are close to zero, which

Table 10.2 Characteristics of the selected distributions for all AC systems.

Descriptives for VAV WLCC model Model	Mean	Variance	Skewness	Kurtosis
Sample	3824.859	107226.2	0.14612	3.00017
1 – Gamma	3824.859	107312.9	0.17129	3.04401
Descriptives for VRF WLCC model Model	Mean	Variance	Skewness	Kurtosis
Sample	3539.283	41990.82	0.1635	2.99374
1 – Random walk	3539.283	41987.63	0.17349	3.05014
Descriptives for fan-coiled WLCC model Model	Mean	Variance	Skewness	Kurtosis
Sample	3250.953	39889.1	0.10615	2.77496
1 – Johnson SB	3252.941	39889	0.1	−0.219
Descriptives for chilled ceiling WLCC model Model	Mean	Variance	Skewness	Kurtosis
Sample	1507.425	9731.299	1.17E-03	2.64078
1 – Johnson SB	1506.844	9731	0.0012	−0.355

indicates that the selected distribution for representing maintenance cost is a good match. Based on the above evidence and deduction, it appears that the selected distributions provide a good representation for the PV values. To validate this deduction, a goodness-of-fit test is carried out to assess formally the quality of the selected distribution representation. The results of this test are shown in Table 10.3. Since the modified statistic test is significantly lower than the critical values for level of significance shown in Table 10.2, the selected distribution cannot be rejected at all alpha levels indicated in the table. Hence, there is no reason to believe, based on the above evidence and tests, that the selected distribution does not provide a good fit for PV values of each of the investigated AC systems.

Figure 10.3 shows PV percentiles comparison of the floor AC systems. Using the median as a benchmark for comparison, the VAV system is found to be more expensive to install and run over the life-cycle, whereas the chilled ceiling system is found to be the most cost-effective option to acquire and operate. The rationale behind this can be attributed to factors that are discussed by Cassidy (2000). Fan-coil based systems are extremely adaptable and, as a proven technology, their operational reliability and stability are well documented. The main disadvantage with these AC systems is related to their running costs and maintenance requirements, particularly where the density of the fan-coil units is high (Nicholls 1999).

Chilled ceilings and passive chilled beams only provide cooling; separate heating and ventilation systems are required and must taken into consideration when whole life cost analysis is carried out. Maintenance of these systems is minimal, limited to cleaning the coil of dust. But these AC systems have a higher initial capital cost and require a high performance façade to control solar load at the building perimeter. This will not affect the cost of AC systems but does affect the total building cost (Green & Wall 1999).

Assessing the economic viability of AC systems over their life expectancy is difficult to establish with certainty. Hence, it is very difficult to project future cash flows as single point cost forecasts. Monte Carlo simulation provides a tool which assists cost analysts to take into consideration the variability and uncertainty in the modelling of the input parameters. In this study the input parameters in equations (1) and (2) are considered random variables with triangular distributions (although other distributions could be assessed for their goodness-of-fit representation if data is sufficiently available). Monte Carlo simulation allows the entire distribution of PV of each investigated AC system to be analysed, and informed decisions can be taken regarding the viability of each system. Hence, instead of computing a single value of PV for each AC system under study, the simulation provides a probability distribution of PV values for each system. The results generated from this simulation model presenting PV by equations (1) and (2) are the expected values and standard deviations of PV. The PV distribution of each AC system provides an indication of the economic viability of each AC system, and the standard deviations provide information to decision makers about the variability or uncertainty associated with each AC system.

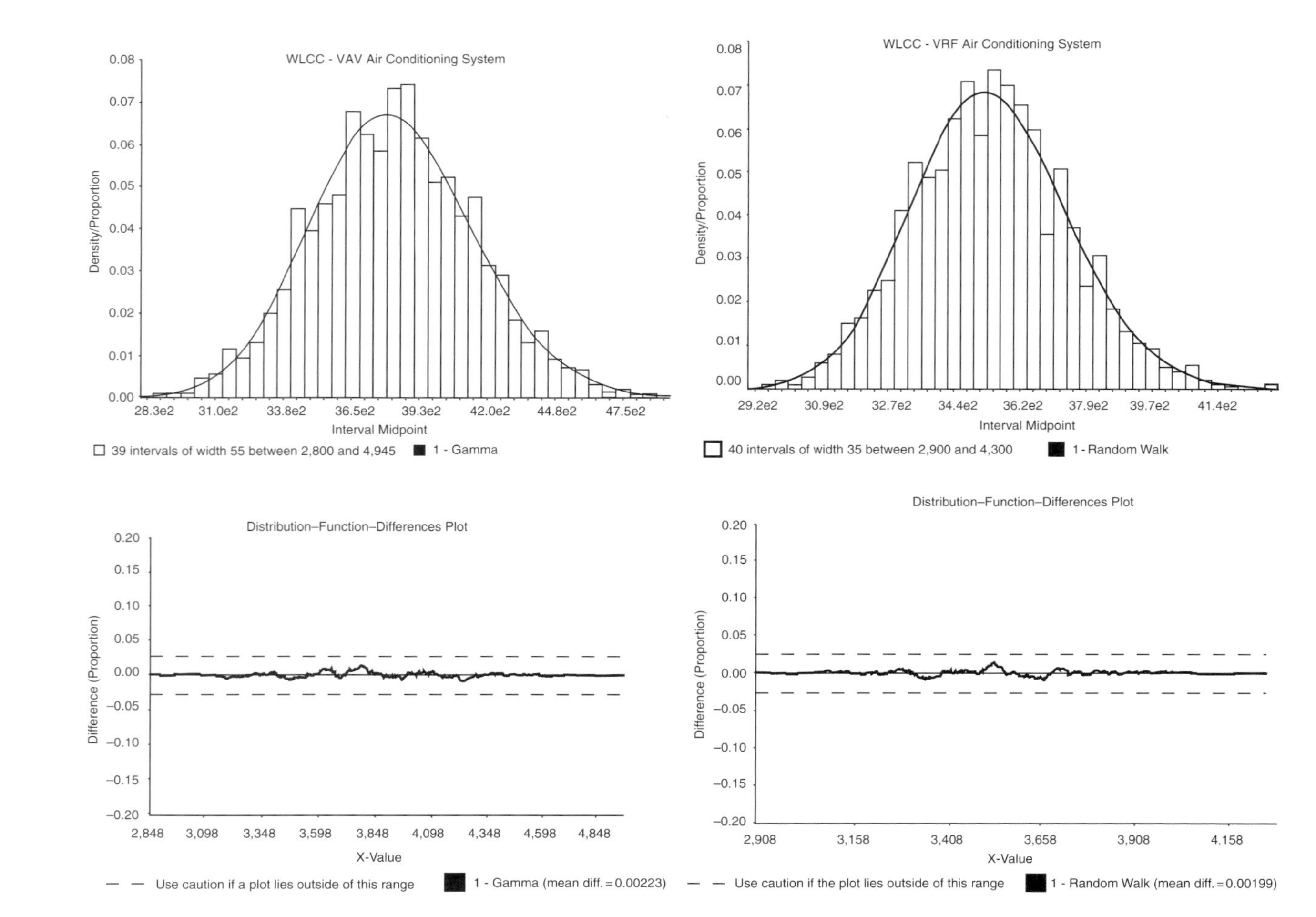

Fig. 10.2 Probability distribution functions and distribution function difference plot for all air-conditioning systems.

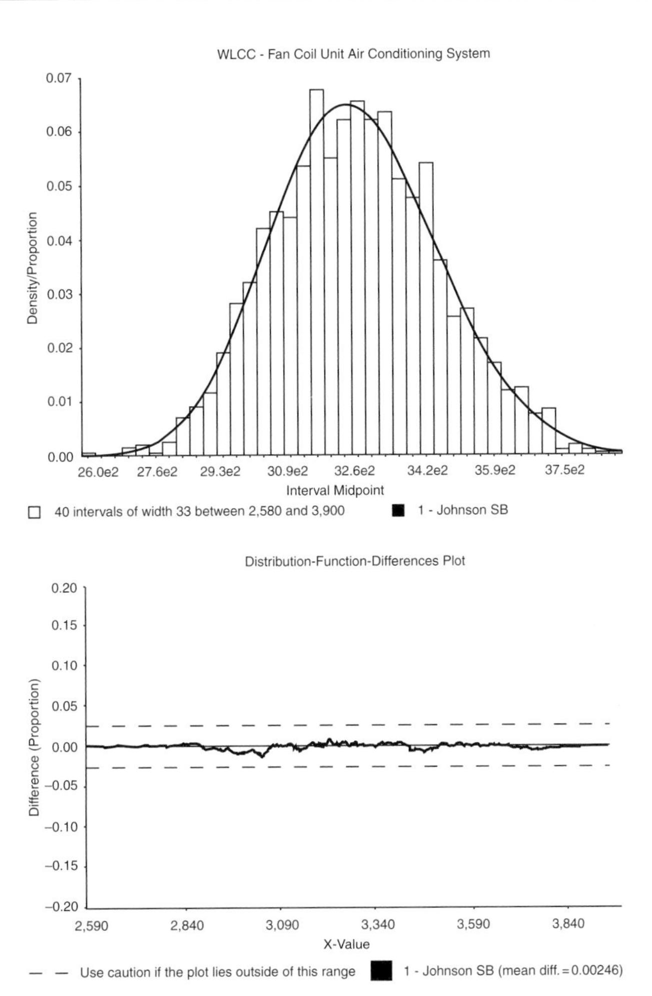

Fig. 10.2 Probability distribution functions and distribution function difference plot for all air-conditioning systems. (Contd)

Table 10.3 Results from Anderson-Darling goodness-of-fit test.

WLCC model	Distribution	Anderson test value	Critical values for level of significance (alpha)					
			0.25	0.1	0.05	0.025	0.01	0.005
VAV system	Gamma	0.26092	0.47	0.63	0.752	0.873	1.04	1.159
VRF system	Random walk	0.21682	1.248	1.93	2.492	3.07	3.86	4.5
Fan-coil system	Johnson SB	0.305	1.248	1.93	2.492	3.07	3.86	4.5
Chilled ceiling system	Johnson SB	0.489	1.248	1.93	2.492	3.07	3.86	4.5

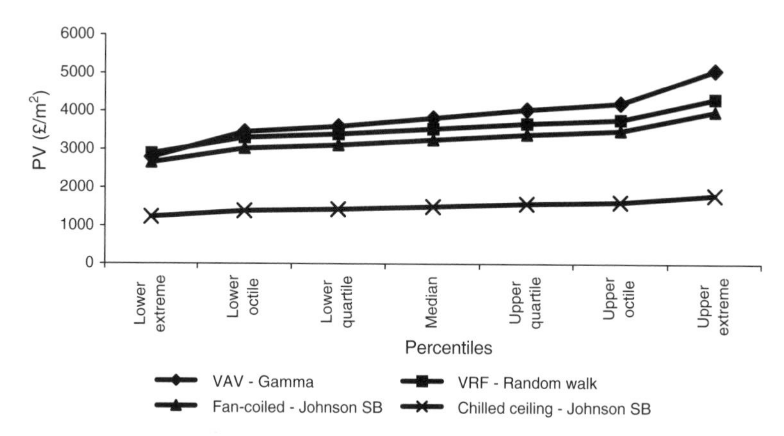

Fig. 10.3 Percentiles comparison of the four systems.

10.5 Summary

This chapter presented a stochastic model which analyses the economic viability of air-conditioning systems. The data used in the study is obtained from published literature. A stochastic software package is utilised to analyse data. The approach incorporated the notion that the whole life cost of these systems depends on several random latent variables. Hence, the input variables to the model are treated as random quantities. The analysis shows that gamma, random walk and Johnson SB distributions best represent the four investigated systems. The chilled ceiling and beam option is found to be cost-effective over the whole life study period. The resulting probability distributions enable all stakeholders to define costs that can be associated with prescribed levels of confidence, thus allowing a quantification of the exposure to budget overruns and financial risk in acquiring AC systems. This is essential for preparing annual management plans and budgets.

References

AbouRizk, S.M., Halpin, D. & Wilson, J.R. (1994) Fitting beta distributions based on sample data. *Journal of Construction Engineering and Management*, **120**(2), 288–305.

Back, W.E., Boles, W.W. & Fry, G.T. (2000) Defining triangular probability distributions from historical cost data. *Journal of Construction Engineering and Management*, **126**(1), 29–37.

Brandon, P.S. (1978) A framework for cost exploration and strategic cost planning in design. *Chartered Surveyor Building and Quantity Surveying Quarterly*, **5**(4), 60–3.

Brown, H.W. (1987) Predicting the elemental allocation of building costs by simulation with special reference to the costs of building services elements. In *Building Cost Modelling and Computers* (ed. P.S. Brandon). E. & F.N. Spon, London.

Building Services (2002) Market watch mechanical and electrical. *Building Services Journal*, **24**(2), 16–17.

Bull, J.W. (1993) *Life-cycle Costing for Construction*. Blackie Academic and Professional, London.

Burgoyne, M. (1996) Current thinking. *Building Services Journal*, **18**(3), 10–11.

Cassidy, T. (2000) Services whole life costs: air conditioning. *Building Magazine*, 4 February.

Chau, K.W. (1995) The validity of triangular distribution assumption in Monte Carlo simulation of construction costs: empirical evidence from Hong Kong. *Construction Management and Economics*, **13**(1), 15–21.

Fente, J., Schexnayder, A. & Knutsom, K. (2000) Defining a probability distribution function for construction simulation. *Journal of Construction Engineering and Management*, **126**(3), 234–41.

Green, M. & Wall, B. (1999) Cost model. *Building Journal*, **3**, 66–72.

Law, A. & Kelton, D. (1991) *Simulation Modelling and Analysis*. McGraw-Hill International Editions, New York.

Nicholls, R. (1999) *Heating Ventilation and Air Conditioning*. Interface Publishing, Manchester.

Swaffield, L. & Pasquire, C. (1999) Examination of relationship between building form and function, and the cost of mechanical and electrical services. *Construction Management and Economics Journal*, **17**(4), 483–92.

Wall, D. (1997) Distribution and correlations in Monte Carlo simulation. *Construction Management and Economics*, **15**, 241–58.

Part III
Whole Life-cycle Costing: Construction and Occupancy Stages

11 Whole Life Risk and Risk Responses at Construction Stage

11.1 Introduction

In Chapter 9 of this book, we considered some of the risks that could arise during the design stage of the project. Identification of these risks is important as this helps to minimise some of the risks that could arise during the construction stage. Risk in construction, however, has always been recognised as a particular problem, not so much with respect to identification, but mitigation. In this chapter we will consider the problem of WLCC risk in the construction phase and how the techniques discussed earlier in the book can be used to identify these risks effectively.

11.2 WLCC at the construction stage

Although much of the recent advancement of WLCC has focused on the role of the construction client, and emphasises the gains to be made on their behalf, whole life costing is a beneficial tool for organisations involved at all stages in the life of the project (WLCF 2003). Although WLCC analysis has in most cases been used solely during the design stage, we shall look later in this book at how WLCC can be applied during the occupancy stage. Firstly, however, how can WLCC be best utilised during the construction phase? At the construction phase of the project, there are three broad applications of WLCC that should be considered (Ashworth & Hogg 2000), as follows.

Method of construction

The method of construction that the contractor chooses to employ on the project can have a major influence on the timing of cash flows and hence the time value of such payments. An appreciation of the time value of money is one of the underlying principles of WLCC; therefore it is of paramount importance that this is taken into consideration.

Selection of components

Unless the design team has specifically prescribed certain elements, the contractor determines the actual selection of many components. In many circumstances, the contractor may select a component or components that comply with the specification of the design but have a different effect on WLCC. For example, two different types of cladding may meet the performance specification laid down by the design team, but one cladding may reduce the energy efficiency of the building. This will subsequently increase WLCC.

Monitoring capital WLCC assumptions

During the course of the construction period, and particularly on large civil/building engineering projects, the assumptions that were used to model WLCC during the design phase should be monitored. If it transpires that changes in the major cost assumptions have been made, then the WLCC analyst must recalculate the forecasts to take into account the changes in the original assumptions. One of the most likely variations in the assumptions is labour cost. The construction industry encompasses a very volatile cost base of labour, and the current lack of skilled labour can significantly increase the original estimates.

11.3 WLCC risk during the construction stage

Risk in construction has been the focus of attention mainly as a result of the cost and time overruns associated with many projects. Miller and Lessard (2001) consider construction risk in the context of 'completion risks' and these are defined as 'the difficulties that clients, principal contractors, and subcontractors may face in the actual building phase of the project'. This emphasis of risk on the organisations responsible for the construction phase is considered further in Baloi and Price (2003) who look at construction risk, not as a physical part of the process but rather as a characteristic of the organisation. Their research shows that most of the risks identified in projects are sourced at the construction phase rather than during the operational phases. It is important therefore that the risks during the construction phase are identified, as this stage forms an important part of the WLCC.

11.4 Typology of WLCC risk at the construction stage

The literature has revealed many efforts to classify risk during the construction phase of the project. At the most basic level, construction risk can be characterised by three risk parameters (CIRIA 1996):

- Risk to activity (e.g. delay in completion, increased cost due to labour shortages, failure in plant and equipment supply)

- Risk to health and safety (e.g. personal injury, downtime due to prohibitions of works notice, discovery of dangerous chemicals on site)
- Risk to environment (e.g. pollution caused by construction process, noise resulting from construction process).

This simplified characterisation of risk though is unhelpful; risk is perceived more as a problem associated with ownership of risk rather than anything else. Kartam and Kartam (2001) considered the risks prevalent at the construction stage in terms of risk allocation. In a questionnaire they sought to identify risks that could occur during the construction stage, and who would be responsible for these risks. They defined construction risk into two groups, risk allocation and risk significance. The risk allocation was subdivided into four principle groupings: risks that were carried by the contractor, risks carried by the client, risks that were shared and finally risks that had no clear ownership. In Table 11.1, the risks identified are shown with possible responses and mitigating actions for each risk.

11.4.1 Contractor risks

In most construction contracts, the contractor bears most of the risk in the construction stage; therefore the contractor's liability through the whole life of the building is quite low. However, in PFI/PPP projects the situation is different; the contractor takes on more of the risk during the occupancy stage and therefore the risk through the whole life-cycle is increased. The principal risk transfers in PFI are generally (Highways Agency 2003):

- Construction and operational cost overruns
- Delay in delivery of the service
- Design of the underlying asset not delivering the agreed service
- Changes of law, including tax law changes, which impose additional or increased costs on the operator (other than any change of law which discriminates against private sector operators).

In more traditional contracts, however, the risk emphasis is mainly on the contractor and the principal risks would include the following.

Labour, material and plant availability

Availability and quality of labour, material and plant can have major implications for WLCC and project success during the construction stage. In the current climate, availability of skilled labour is a major risk to the project, with many high profile projects suffering severe cost overruns as a result. Material and plant availability are less risky provided that adequate planning and ordering are provided for.

Table 11.1 Typology of risk allocations in the construction phase.

Risk allocation	Risks identification	Risk responses/reduction approaches
Contractor	Labour, material and plant availability	Maintain good relationships with suppliers, avoid late payments
	Industrial disruption	Ensure parity with national pay scales/working hours
	Labour and equipment productivity	Regular formal and informal training
	Subcontractor coordination	Establish formal channels of communication with subcontractors
	Health and safety	Compliance with the statutory health and safety plan/file, etc.
	Workmanship	Regular inspection and training
	Project programme accuracy	Regular review and revisions of the programme
	Competency of the contractor	Use selective tendering
	Defective materials and components	Use certificated components and materials (i.e. BBA)
	Different site conditions	Thorough survey before construction begins
	Actual quantities of work	Thorough review and revision if necessary of the bill of quantities
	Adverse weather conditions	Monitoring of forecasts on a day-by-day basis
	Price inflation	Incorporate a contingency value in the capital cost
Client	Low quality of advice	Seek contractual advice from more than one source
	Income/benefits from project lower than anticipated	Ensure that market research is thorough and results validated
	Delays in completion	Monitoring of progress with construction and design team
	Unforeseen project costs	Allocate contingency finance
Shared	Variations (agreed)	Avoid complex deviation from the original design
	Acts of God	
	Force majeur	
	War threats	
	Financial failure	Monitoring of cash flows and institutional finances
No clear allocation	Site access	
	Defective design	Close client–design team cooperation
	Government legislation	
	Third party delays	Ensure clear channels of communication for all stakeholders

Industrial disruption

The risk of strike action by site operatives will naturally have consequences for WLCC and progress. The existence of national pay bargaining and improved union relations has helped to reduce this risk, although poor management relations on site could easily lead to problems.

Labour and equipment productivity

Although securing the required labour and plant is vital to the project, ensuring that this is of sufficient quality is equally important. Formal and informal training can help to increase labour productivity during the contract and similarly, adequate maintenance regimes must be in place to ensure that the plant operates efficiently. Failure to account for this will lead to delays in the project.

Subcontractor coordination

This risk has been particularly identified with respect to specialist subcontractors. Failures to coordinate the work plan correctly can lead to critical delays and WLCC overruns. Specialist subcontractor work can account for a high percentage of the overall capital cost; therefore the effect on the overall project is of a greater magnitude.

Health and safety

Compliance with health and safety legislation is not only a legal requirement, but also a vital part of risk reduction during construction. Formulating a strong health and safety regime will help the contractor to comply with the law whilst reducing the risk of accidents and costs of stoppages in work due to non-compliance.

Workmanship

The risks associated with poor workmanship can manifest not only during the construction phase but also during the occupational phase of the building. At construction phase, poor workmanship can lead to reworking and project cost increases, whilst during the occupation phase problems can arise with early component failure. Workmanship directly affects WLCC as this is strongly correlated to the service life of elements and the building.

Project programme accuracy

If the planner has failed to account for potential time lags and unforeseen delays in material delivery, plant and labour availability, deviation from the project critical path will lead to expensive cost overruns and project completion delays.

Competency of the contractor

The ability of the contractor to complete the works to the agreed contract is a risk that can be reduced by ensuring that adequate research has been conducted on the contractor's past work history and experience of similar projects. An incompetent contractor can increase project WLCC budget.

Defective materials and components

Ensuring that materials and components of a suitable standard and quality are used is essential in ensuring that the building functions effectively during the occupancy stage. This risk is more apparent in PFI/PPP schemes where the consortium will be focused on selecting materials with a suitable life-cycle. Components that fail earlier than expected can cause expensive response maintenance costs and lead to increased WLCC. An interesting risk trade-off also occurs here when considering PFI/PPP – the PFI consortium will want to utilise components that last no longer than the contract service life, whereas the client will seek to specify components that are less likely to fail over the period following the contract service life.

Different site conditions

Ensuring that a thorough investigation of the site conditions is performed at the outset can reduce the risk of different site conditions to those anticipated interrupting the progress of the construction period.

Actual quantities of work

Providing that the bill of quantities has been drawn up correctly, this risk is low. However, should variations to the project be required, differences between actual and planned work could lead to cost and time overruns.

Adverse weather conditions

The contractor should take steps to ensure that regular updates of weather conditions are provided so that the work plan can be altered should adverse weather conditions occur that are likely to interrupt the work programme. Adverse weather could also impact on WLCC through damage to components and elements during the construction phase. Inadequately protected workings are more likely to fail earlier than expected.

Price inflation

This risk is low, as price inflation cannot be foreseen to be substantial during the contract period.

11.4.2 Client risks

The risks to the client during the construction stage are not of the same magnitude as the risks to the contractor. The risks to the client are more apparent during the occupancy stage of the building, when issues related to financing rentals and development potential need to be considered. In PFI/PPP projects, the client risk (i.e. the public sector) is assumed to be reduced, although many would argue that this is not the case. The main risks to the client during the construction phase include the following.

Low quality of advice

If the advice provided to the client is of a poor standard this could lead to inappropriate management tools used during the construction phase, inappropriate selection of the procurement route, failure to secure the desired level of quality of building and a poor value for money building. This will inevitably lead to a higher WLCC, as the cost optimisation process will not have been completed.

Income/benefits from project lower than anticipated

The potential failure of the building to realise income during the occupancy stage can carry considerable risk, particularly with regard to project finance arrangements. Many WLCC calculations that consider revenues to offset costs will also be severely affected by any materialisation in this risk. Developers will be particularly keen here to ensure this risk is minimised.

Delays in completion

These risks are similar to those mentioned earlier – delays that could be caused by disruption, industrial disputes, poor quality work and financial problems, which can again lead to increased WLCC.

Unforeseen project costs

This primarily includes variations to the original design but can also include costs during the occupancy phase, as a direct result of costs incurred at construction such as defective design, specification failures and defective construction. These unforeseen project costs must be reappraised in the WLCC model.

Construction stage WLCC risk can be characterised not only by allocation, but also by identification, expanding on the three-risk parameter characterisation discussed at the start of this section. In the example shown in Fig. 11.1, construction WLCC risk is defined as the coalescence of technical, financial, organisational and environmental risk. By defining the risks in this way, we can consider the inter-relationships of risk between the stakeholders.

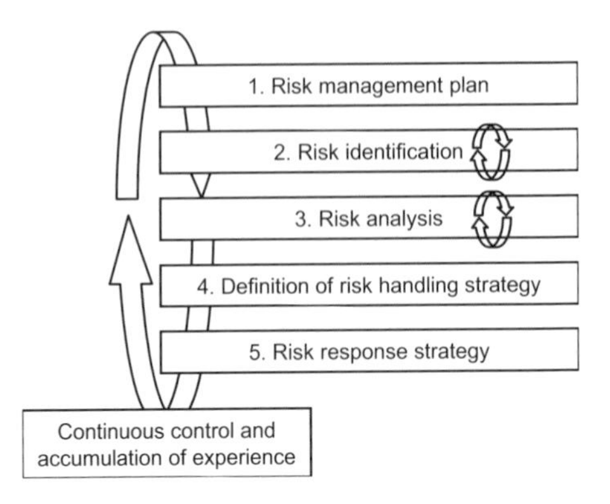

Fig. 11.1 Systematic project risk management process.

The diagram includes many of the risk factors identified in Kartam and Kartam (2001) along with several others, which are considered to be of importance.

11.4.3 Technical risk

Technical risk deals with the esoteric aspects of the project. This includes process, engineering, manufacturing, materials and information systems cost. Complex projects can present serious problems to the constructor. Potential difficulties may be overlooked during the tendering process, which could lead to schedule and cost problems on the project. Technical risk must be identified during tendering, and consideration given to procuring specialised equipment or employing qualified professionals to perform or manage the work. Technical risk exposure should in theory be more manageable than perhaps the other risk umbrellas, the rationale being that good research, high-quality work and correct specification should reduce the risk of increased WLCC.

11.4.4 Financial, economic and political risk

Financial risk encompasses the risks associated with the financial operations and management of the project. These risks could include cash flow risk, failures in securing project funding at key stages in the project, incorrect valuation and stage payment procedures, large project retentions, etc. Economic and political risk considers some of the risks that can impact on the project from outside the influence of the project stakeholders. The risks that could be categorised under this heading include inflation risk, changes in existing legislation, changes in government policy that have direct consequences for the project, incorrect estimation procedures and failures in the contractual arrangements. It is difficult to devise effective risk mitigation strategies for some of these risks as the magnitude and effects can be similarly difficult to foresee.

11.4.5 Organisational risk

On the strategic level, organisational risk can be defined as the degree to which the organisation is capable of responding to and carrying out the changes required in a project. On a more detailed level, organisational risk considers the adverse effects of failures in labour and the subsequent consequences of this on the project. Such instances include the availability of labour, competency of the various teams involved in the construction of the building, capabilities of materials suppliers and ability of financing institutions on persons to ensure adequate cash flow during the construction period.

11.4.6 Environmental risk

The construction industry poses a significant threat to the UK environment. Construction companies are responsible for more pollution incidents than any other industry sector. In other sectors incidents are decreasing; in construction they increased by 20% between 1996 and 1999 (RPA 2000). One reason for the industry's lack of progress in reducing pollution is its apparently limited business impact. But few companies realise the true costs of pollution incidents. Information on pollution costs through the whole life-cycle, and not just during construction, can help organisations to ascertain how best to address the risks and reduce the cost impacts. Environmental risk also considers the risk associated with the influence of the external environment on the project (such as weather) and of environmental legislation which can affect the location of the construction project, construction method and working hours, for example.

11.5 Tools for allocating WLCC construction risk

Risk allocation is, as we have discussed earlier in this chapter, one of the most important aspects of WLCC risk during the construction phase. To ensure that the risks are allocated in the optimum approach, a variety of techniques can be used. The following points should be considered in WLCC risk allocation (Pickavance 2000):

- Risk is generally assumed to be subject to the following principles: risks should be allocated to the party best able to control them, i.e. the party which is best able to forestall the risk or to minimise its effects if it occurs
- Risk allocation should encourage risk management by the party best able to manage the risk. For example, a management contractor should shoulder the risk of delay caused by works contractors
- The party that does not assume primary responsibility for risk should nevertheless be motivated to manage the consequence of the risk if it materialises.

The core strategy for risk allocation should be developed during the earliest stages of the procurement process, as this is vital in ensuring that the parties most able to manage the risk are identified. This is not to say, however, that risk allocation cannot occur during the construction phase. The risk allocation process can be seen as two stage: stage one involves the qualitative risk allocation phase that utilises the techniques described in Chapter 5 to identify what types of risk should be allocated to whom. The second stage involves the use of quantitative approaches that can be used to determine how much risk should be allocated. Simulation modelling can be used, for example, to derive the probability distributions that best describe the risk allocation. However, other tools for risk allocation do exist and one of the most common types is the contract governing each participant's project responsibilities. The contract can help to allocate the risks during the construction phase by identifying:

- The responsibilities of the client
- The responsibilities of the contractor
- The stages and milestones by which compliance with the responsibilities can be measured
- Mechanisms for ensuring the correct inspection, measurement and payment of completed works.

Without any formal arrangements for risk allocation, disputes during the construction phase can lead to a critical failure in progress. Furthermore, significant delays could lead to project finance difficulties.

11.6 Significance of WLCC risk at the construction stage

Having identified the risks and allocated them accordingly, the next stage should involve identifying the significance of each risk on the construction of the building. These weightings are not generic and will depend upon the project under consideration. For example, a prestigious high-quality building project will have a larger emphasis upon quality and therefore the risk associated with workmanship and component quality will carry a higher weight. A building that is constructed in an area with a shortage of skilled labour will carry a higher risk of ensuring that the project is adequately staffed. Some of the techniques that were explained in Chapters 4 and 5 of the book can be utilised for these purposes.

11.7 Monitoring WLCC risk throughout the construction programme

The risks that have been identified will change as the construction period moves on and the work nears completion. It is important therefore to ensure that the WLCC risk management process throughout this time is iterative, so that new risks that may occur are quickly identified and allocated. This technique is commonly referred to as the observational approach – monitoring the symptoms of risk and taking action to keep the anticipated outcome within acceptable limits.

A variation on the theme of the observational approach is systematic construction project risk management referred to by Kähkönen and Huovila (2003), which also emphasises the requirement for continual monitoring throughout the WLCC process. The technique includes five phases, starting from the definition of risk management plan for the project in question and, finally, reaching the response-planning phase (Fig. 11.1). The five phases form only a framework for systematic project risk management practice, which is supposed to employ continuous response control and gathering of experiences and know-how.

11.8 Construction WLCC risk responses

Kartam and Kartam (2001) identified two kinds of responses to risks that have been identified during the construction phase, these being preventative action and mitigative action. Preventative actions can reduce the risks at an early stage of the construction period, but with the possible side effect of leading to an artificially high bid during the tender stage. Mitigative actions in contracts are aimed at remedial action with the intention of minimising the risks as far as possible. The authors identified seven preventative methods:

- Utilise quantitative risk analysis techniques for accurate time estimation
- Use of subjective expert judgement to devise the most appropriate programme of works
- Produce a project schedule based upon obtaining updated project information
- Plan alternative methods/options as stand-by
- Consciously adjust for bias and add risk premium to time estimation
- Transfer or share risk to/with other parties
- Refer to previous and ongoing similar projects to produce accurate works programme.

Although these techniques are widely available, along with many software packages in which to generate programmes of work, most delays do occur during the construction phase and as such, mitigative actions are required. The most common mitigative responses identified include the following:

- Increase manpower and/or equipment
- Increase working hours
- Change or modify the construction method
- Change or modify the sequence of work by overlapping activities
- Coordinate closely with subcontractors
- Close supervision of subordinates to ensure minimisation of abortive work.

11.9 Summary

This chapter has looked at an aspect of WLCC that is rarely discussed, the application of WLCC during the construction stage. It has been mentioned on many occasions throughout this book that WLCC is a dynamic process; it therefore starts at the birth of the project and ends when the building is no longer functional. The construction process forms an important part of the whole life; it is where the design is realised. However, the risks during the construction process can have severe consequences on the WLCC of the building, and these risks must be taken into account both before and during the phase of construction. Developing and realising an effective risk management process throughout the construction phase can help to reduce the WLCC risk. This is important in ensuring that uncertainties are quantified and that risks can be handled effectively if and when they occur.

References

Ashworth, A. & Hogg, K. (2000) *Added Value in Design and Construction*. Longman Publishing, London.

Baloi, D. & Price, A.D.F. (2003) Modelling global risk factors affecting construction cost performance. *International Journal of Project Management*, **21**(4), 261–9.

CIRIA (1996) *Control of Risk: A guide to the systematic management of risk from construction*. Special Publication 125, Construction Industry Research and Information Association, London.

Highways Agency (2003) Risk transfer and value for money. In *Design Build Finance Operate (DBFO) – Value in Roads*. http://www.highways.gov.uk/roads/dbfo/value_in_roads/031.htm

Kähkönen, K. & Huovila, P. (2003) *Systematic Risk Management in Construction Projects*. http://web.bham.ac.uk/d.j.crook/lean/iglc4/huovila/risk2.htm

Kartam, N.A. & Kartam, S.A. (2001) Risk and its management in the Kuwaiti construction industry: a contractor's perspective. *International Journal of Project Management*, **19**(6), 325–35.

Miller, R. & Lessard, D. (2001) Understanding and managing risks in large engineering projects. *International Journal of Project Management*, **19**(8), 437–43.

Pickavance, K. (2000) Delay and disruption in construction contracts. *Building & Construction Law*, **17**, 1–660.

RPA (2000) *Guidance on the Costing of Environmental Pollution from Construction*. Construction Industry Research and Information Association (CIRIA), London.

WLCF (2003) *Whole Life Cost Forum, About Whole Life Costs*. http://www.wlcf.org.uk/WhatAreWLCs.htm

12 Whole Life Risk and Risk Responses at Operational Stage

12.1 Introduction

The previous chapter gave an overview of some of the techniques and procedures that are useful in modelling WLCC risk during the construction phase of the building. As with the construction phase, however, we need to appreciate the risks that are inherent during the operational phase. It should be noted that operational risk encompasses events with very differing frequencies, and possibly patterns of occurrence and severities, to that of construction risk. It is important to understand and gain awareness of these risks as this can provide a good deal of information when making future investment decisions in the capital stock. In this chapter we shall investigate where the likely sources of risk are, and what responses should be considered when dealing with this risk. The emphasis in this chapter is on the practicalities of assessing operational risk in WLCC by demonstrating, with examples, the likely risks, consequences and responses. This chapter provides guidance on which of the procedures explained in the early chapters of this book are most appropriate for operational stage WLCC risk analysis. This will then provide the necessary information with which to build an operational stage WLCC model, which will be covered in Chapter 13.

12.2 WLCC risks at operation stage

The study of operational risk is a new concept in WLCC and is rarely referred to in the literature. It is, however, an important facet of the financial management of the building during the occupancy stage. Any future investments in the existing stock such as renewal and replacement of building elements, review of existing facilities management provision and financial credit agreements and future demands on energy supply all have associated risk profiles. For example, consider an air-conditioning unit. Over the life of the building, there is a risk that the element will cease to operate effectively or will even fail. When planning budgets these risks need to be considered so that funds are available to replace these elements should it be required. This is particularly important in critical structures such as hospitals and schools where uninterrupted service of the building is a prerequisite.

In earlier chapters of this book, reference has been made to the WLCC ratio: for every £1 spent on capital, £50 is spent on maintenance and £200 is spent on operational costs. It can therefore be appreciated that with such a heavy emphasis upon operation costs throughout the whole life-cycle, identifying the risks inherent to these costs is important.

12.3 Identifying the WLCC operational risks

In large buildings, operational WLCC risk identification is a complex task, and should not be underestimated. During the operational stages of a building, risk management normally focuses more on the functions of the building (i.e. service delivery) rather than costs. There is, however, little to be gained from failing to continue the WLCC risk identification procedures through the operation stage, as this is where the WLCC really begins to accumulate.

During the operational phase of the building we must consider the separate cost centres of maintenance, energy, facilities management and finance costs. Each of these operational cost centres can then be broken down further to identify the subcomponents of each cost centre. After completing this assessment, the analyst should then proceed to identify any particular risks to each cost component and what actions can be taken in response to the identified risk. To achieve this, some of the qualitative techniques discussed in Chapter 5 could be used, such as risk registers and risk matrices. Figure 12.1 demonstrates the process of operational WLCC risk identification. As the diagram shows, the processes are iterative and should be continual through the whole life of the building. As the functions of the building evolve, more costs will be generated, and the risks to these costs will affect the WLCC. In Chapter 13,

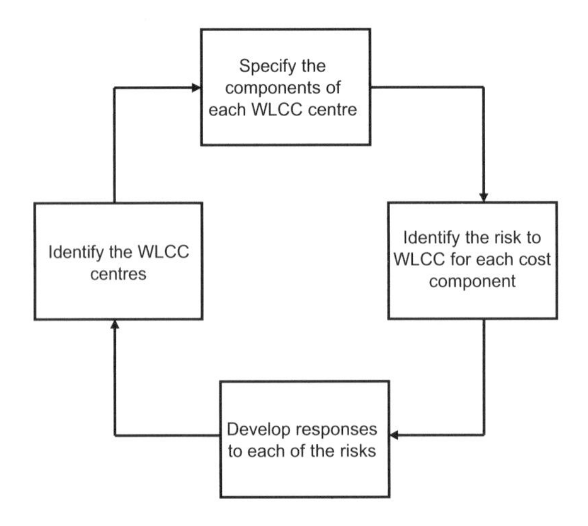

Fig. 12.1 The process of operational WLCC risk identification.

we will see how this process should run concurrently with the operational stage WLCC process, to ensure that the model is updated regularly and takes into account recently identified risks. This procedure and the cost elements to be considered will be discussed in more detail in this chapter.

An alternative approach to identifying operational risks, which would not involve such a detailed analysis, could be by means of establishing a set of performance requirements for the building (sometimes these performance requirements are specified by the client during the briefing stage so these could be used as a basis). These requirements should be monitored over the whole life of the building. It may be the case that these performance requirements were set by the client during the briefing stage of the project – if so these could provide the basis of operational risk identification. The analyst could then assign risks to the achievement of each of these performance targets along with possible responses to remove, reduce or mitigate the risk. This process is shown in Fig. 12.2. Having selected the procedure demonstrated in Fig. 12.1 for assessing operational phase WLCC risk, the next stage involves identifying the relevant operational cost centres. There is no strict definition of what should be included here, as this depends upon the project specifics. The list provided here is an example and should be tailored to the particular project needs.

12.4 The maintenance cost centre

Maintenance costs in buildings can be described as the expenditure with respect to inspections, surveys and work required for the preservation, repair and replacement (with equivalent contemporary items) of existing building components, plant, services or equipment, preventative maintenance and follow-up remedial work and upkeep of external works. In most cases, the work involving adaptations, improvements and/or alterations is not included as this is more concerned with capital expenditure and replacement.

Maintenance costs can be classified into two distinct groups and a further four subgroups (Holmes 1994). Figure 12.3 shows the various forms of maintenance. Response maintenance can be divided into normal maintenance and emergency maintenance, the terms implying the priority given to the work

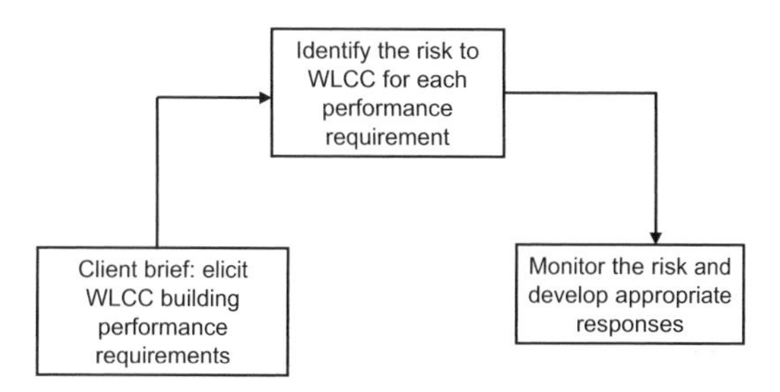

Fig. 12.2 Risk identification through performance indicators.

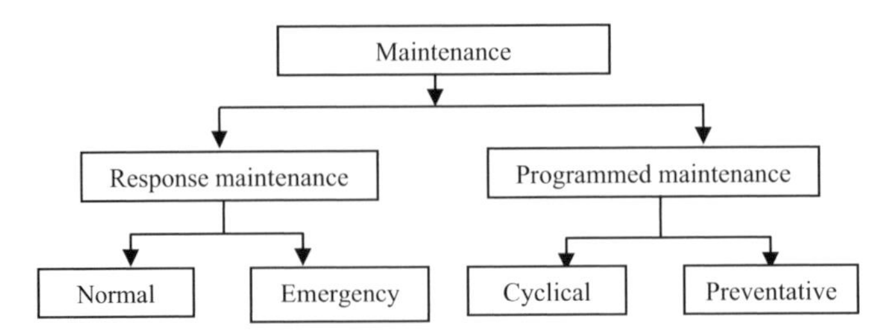

Fig. 12.3 Classification of maintenance regimes.

required. All organisations should draw up an emergency response list to enable problems to be prioritised easily. Response maintenance is usually initiated by the occupier and will be categorised by a number of factors such as type of work required, nature of fault and seriousness. This coding of work gives vital data on response maintenance thus enabling future budgets to be manipulated to take account of future maintenance demands. Programmed maintenance (commonly referred to as planned maintenance) can be further categorised into cyclical and preventative. The former is maintenance carried out on a rotational basis irrespective of the condition of the building element. The latter deals with building elements that have reached their optimum life span, or maintenance under the recommendations of the manufacturer. Although regular, planned maintenance reduces downtime costs, it expends maintenance budgets at a higher rate, but adversely the 'run until it breaks' approach reduces maintenance expenditure but increases downtime loss.

Maintenance provision in buildings is usually categorised into three distinct headings: engineering maintenance costs, building maintenance costs and external works maintenance costs (Fig. 12.4).

12.4.1 Building maintenance WLCC

The building maintenance WLCC centre should represent all maintenance expenditure for the building on walls, roofs, windows, internal floors, ceilings, partitions, etc. The work relates to the internal and external aspects of the building fabric and its components. Items such as roof replacements are not included in this cost centre, since this would represent a modification and hence be included in any capital expenditure calculation (this is considered as a residual cost).

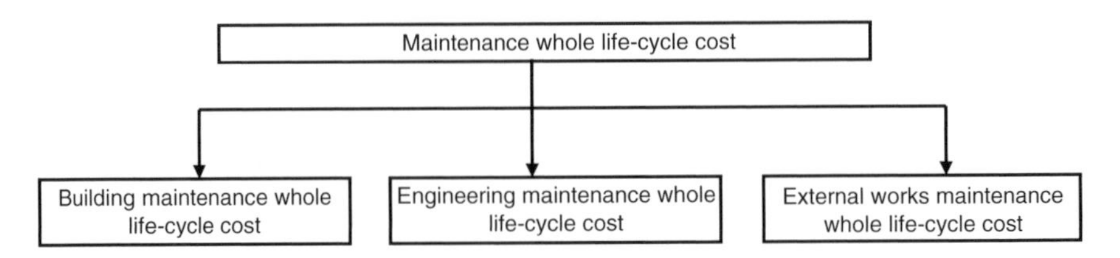

Fig. 12.4 Maintenance provision in buildings.

12.4.2 Engineering maintenance WLCC

The engineering maintenance WLCC centre should include all expenditure on maintenance of calorifiers, generators, security systems (except those included under the facilities management cost centre), boiler plant, electrical installations, equipment, etc. that are internal, i.e. within the confines of the lowest slab level and the inner surface of the building(s). Any roof-mounted equipment, which is associated with internal equipment such as air-conditioning units, is included in this cost centre. Infrastructure associated with the engineering plant that runs external to the building(s) is included, such as pipe-work and distribution cabling, and any building management systems (BMS) are also included.

12.4.3 External works maintenance WLCC

This cost centre relates to the maintenance of grounds, gardens, paths, roads, car parks, street lighting and external sewers, etc., i.e. work that is external to the fabric of the building.

12.5 The energy cost centre

The energy cost centre should include the total annualised cost of all energy supplies to the building. The energy cost centre should be classified by separate cost categories for each of the following:

- Electricity WLCC
- Natural gas WLCC
- Fuel oil WLCC
- Energy from other sources (WLCC).

In general the costs should be derived by billed amounts although in certain circumstances consumption in kW should also be recorded. This would be in the case where the building obtains energy from a combination of different sources where the unit cost is variable. The classification of energy cost centres is shown in Fig. 12.5.

12.6 The facilities management cost centre

Facilities management (FM) can be defined as the practice of coordinating the physical workplace with the people and work of the organisation, integrating the principles of business administration, architecture, and the behavioural and engineering sciences (IFMA undated).

This definition of FM is expanded further to suggest that its purpose is to increase the value of services, either by improving quality of service for the same cost, or delivering the same service for a lower cost – or indeed a combination of both these scenarios (NHS Estates 1996). This definition of FM highlights the emphasis on cost, and the importance that this places particularly on WLCC. FM services may be provided in a variety of ways, ranging from completely in-house service provision through to completely out-sourced services such as in the case of PFI hospitals. The facilities management cost centre can be subdivided into many individual WLCC components. The list of subdivisions shown in Fig. 12.6 and summarised below is not intended to be

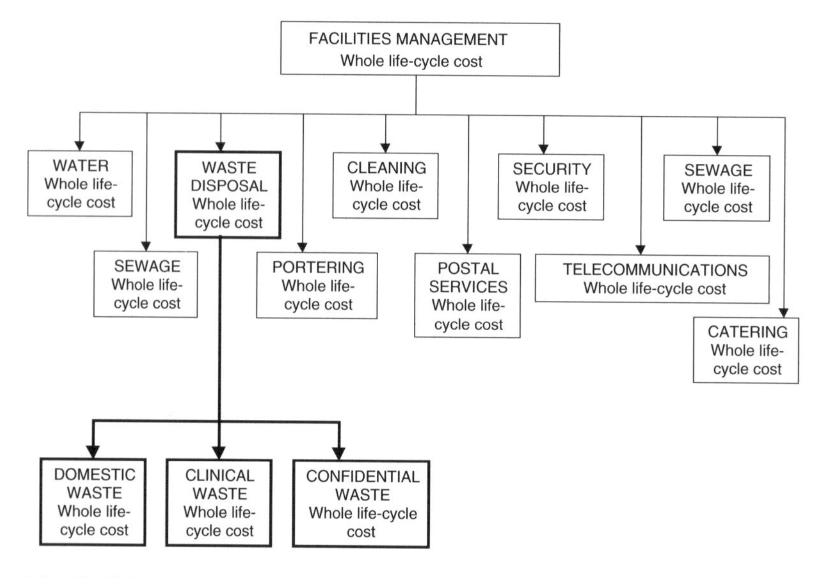

Fig. 12.5 Facilities management provision, with example of the subdivision of the waste disposal cost centre.

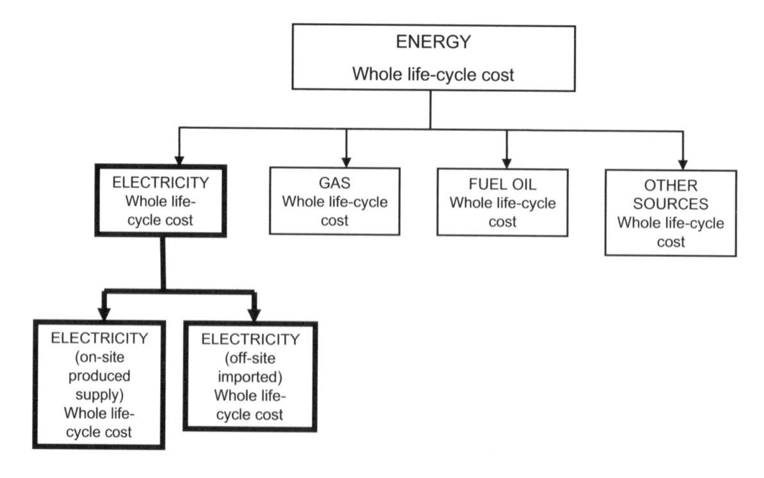

Fig. 12.6 Energy cost distribution in buildings, with example of the subdivision of the electricity cost centre.

exhaustive, but illustrative of the typical FM cost elements that should be included in the risk identification process.

12.6.1 Water cost

This cost relates to the total cost incurred, inclusive of borehole maintenance and operational costs together with billed amounts. Water costs associated with on-site central processing units(s), for example laundry, should also be included in this cost centre. In some cases water and sewage are expressed as one single cost centre (as the sewage is billed as an inclusive amount of water cost); however, this is not recommended here as in certain cases sewage is a cost centre in its own right. If this is not the case, a cost of zero can be returned in the calculation.

12.6.2 Sewage cost

Where this cost is not inclusive of water cost, then sewage costs associated with on-site central processing should be included here. Any specialist sewage removal process (normally associated with clinical/chemical sewage) should also be included. Borehole maintenance works in connection with sewage removal are also to be included so long as this has not been apportioned in the water cost centre.

12.6.3 Waste disposal cost

This cost centre represents the total cost of all waste disposal throughout the building inclusive of the disposal of confidential and sanitary waste and recycling. The full cost of employment of persons in connection with waste disposal should also be included. The cost of disposal is defined as the costs involved in off-site disposal including transport and disposal costs. This cost does not include on-site collection and handling costs associated with bringing waste from the various areas within the building to a central collection point. In certain circumstances (i.e. hospital buildings) this cost centre should be subdivided into specialist waste disposal costs such as domestic, clinical and confidential.

12.6.4 Portering cost

The portering cost centre represents all pay and non-pay costs for the building. Pay costs include all elements related to the cost of employment of directly contracted staff for the purpose of carrying out all portering duties. This includes all departmental portering staff, except out-sourced contract staff. A non-pay cost includes all costs incurred in meeting the portering services for the building. This includes any contract staff costs, training, and equipment and uniform costs associated with the pay costs.

12.6.5 Cleaning cost (non-maintenance related)

The cleaning service cost centre represents total pay and non-pay expenditure on cleaning provision throughout the building. Pay expenditure includes costs associated with salaries and superannuation of all staff, including managers, supervisory staff and administrative functions. It does not include costs associated with the cleaning of leased out space within a building. Non-pay expenditure includes all cost associated with the purchase of materials, consumables, uniforms, equipment and the subsequent maintenance of this equipment.

12.6.6 Postal services cost

The postal cost centre represents all pay and non-pay costs for the entire building. The pay cost should include all costs associated with sorting and distribution for delivering incoming mail and for the collection of outgoing mail inclusive of letters, package and parcel cost. It could also include all costs associated with the internal mail system. Non-pay costs normally include the costs associated with postage and courier charges, equipment, uniforms, consumables, stamps, contract staff and, where appropriate, systems maintenance costs (not including building/engineering) which are not included in the pay cost.

12.6.7 Security service cost

The security cost centre represents all pay and non-pay costs for the entire building. Pay cost should include all elements related to the cost of employment of directly employed staff for the purpose of security services. It should also include all non-security staff seconded to security duties. Non-pay cost includes all equipment, uniforms, consumables, contract staff and system maintenance costs for CCTV and door entry systems. The total of security service pay and non-pay costs is the total costs incurred to supply the security service. This figure does not include any income/costs generated from car parking-related elements.

12.6.8 Telecommunications cost

The telecommunications cost centre represents all pay and non-pay costs for the entire building. Pay cost includes all elements related to the cost of employment of directly employed staff for the purpose of telecommunications. It does not include contract staff and out-sourced services to the trust. Non-pay cost includes the cost associated with telephone, internet and other costs, including the maintenance of these systems.

12.6.9 Catering cost

This cost centre represents the cost of providing all catering and vending services. This should be a net cost inclusive of any income derived from the licensing of catering space as part of a catering contract. The cost centre should also include the cost of food and drinks including preparation, crockery and cutlery.

12.7 Identifying the operational stage WLCC drivers

In order to be able to identify the risks to WLCC for each of the operational cost centres, we need to identify the likely cost drivers. The cost drivers are the characteristics of each cost centre, which have some magnitude of impact upon the WLCC. If a thorough identification of the cost drivers has been carried out, this facilitates an easier approach to WLC risk identification. Table 12.1 identifies some of the possible drivers that influence operational WLCC, although these again will differ according to the particular type of building (IPD 2002).

12.8 Operational stage WLCC: risks and risk responses

Once having identified the operational cost drivers, this information can then be used as the basis for identifying and defining the WLCC risks. Some of the principal WLCC risks that could occur during the occupancy stage include (The Stationery Office 2000):

- Contractors fail to meet performance standards for service delivery
- Main contractor fails to make assets available for use
- Operational costs are more than forecasted
- Operational costs are less than expected
- Assets underpinning service delivery are not properly maintained.

As with the techniques described in Chapter 11, a variety of risk responses are available. Where possible the analyst should try to optimise the risk responses although obtaining the best possible response may in itself increase cost liabilities (i.e. extended warranties). Responses must be clearly defined so that it is possible to assess their effectiveness. If appropriate measurements cannot be collected the analysis will be limited. The emphasis here is on how well risks themselves are identified and managed (not how effective the management of risk processes is). The chart in Table 12.2 shows an example of some risks that could occur during the operational phase of a building. This list is by no means exhaustive but serves as a basis on which to build up a risk profile.

Table 12.1 Operational WLCC centre drivers.

Operational WLCC centre	Example cost drivers
Facilities management: security services	Design specification of fixed equipment, manned security hours, patrol policy, staff per m^2, number of building access points
Facilities management: cleaning	Design specification for cleaning, surface types, necessity for special cleaning (i.e. sterile areas), labour costs
Facilities management: waste disposal	Quantity of waste, type of waste, availability of and cost of specialist contractors if required
Facilities management: water	Building occupancy level, availability of local supplies
Facilities management: telecommunications	Design specification, number of call points, complexity of system, frequency of use, maintenance arrangements
Facilities management: postal services	Frequency of use, cost of local labour, adoption of new technologies in communications
Facilities management: catering cost	Design specification for catering areas, staff subsidy (if applicable), demand for service, vending machine allocations
Facilities management: portering	Frequency of use, availability and cost of labour, distribution requirements, size of building, function of building
Facilities management: sewage	Building occupancy

(Contd)

Table 12.1 Operational WLCC centre drivers.

Operational WLCC centre	Example cost drivers
Maintenance costs: building	Design specification of building components, building size, labour skill and availability, planned maintenance regimes, building exposure, building occupancy level, location
Maintenance costs: engineering	Design specification of M & E components, building size, labour skill and availability, planned maintenance regimes, location
Maintenance costs: external works	Size of external grounds, design specification, labour skill and availability, environmental legislation compliance
Energy: electricity	Building function, demand, competitive supply, location, size
Energy: gas	Building function, demand, competitive supply, location, size
Energy: fuel oil	Building function, demand, competitive supply, location, size
Finance costs	Interest rates, availability of future investment, loan rates, BoE base rate

Table 12.2 Operational stage WLCC risks and risk responses.

Cost centre	Risk identification	Example risk responses
Maintenance WLCC risks	Lack of skilled maintenance labour	Increase training levels, more intensive inspection of works, better recruitment
	Building fabric/component failure	Ensure effective maintenance regimes, regular condition assessments
	Poor quality maintenance regime	Audit existing regimes
	Increased demands on M & E equipment	Ensure that the existing infrastructure is suitable – regular condition assessments
	Unexpected plant and equipment obsolescence (maintenance costs increase)	Extend warranties on critical equipment (a what-if scenario analysis could be conducted to assess the trade-off of increased cost through warranties versus increased maintenance cost through obsolescence)
Energy WLCC risks	Increase in unit cost of energy	Continual monitoring of market prices
	Increase in demand for energy supply	Ensure efficiency measures are working correctly
	Non-competitive pricing/quotes	Continual monitoring of market prices, benchmarking
	Failure in energy supply	Ensure back-up systems are working regularly
	Failure in energy efficiency measures	Staff training

(Contd)

Table 12.2 Operational stage WLCC risks and risk responses.

Facilities management WLCC risks	Dependency on key suppliers (increased unit cost)	Use of competitive tendering for larger contracts
	Failure of supplier to meet agreed operational standard	
	Insufficient purchasing power	Use of buying consortiums
	Theft of assets from within the building and other security failures	Ensure adequate provision of staff to building occupancy ratio, regular asset auditing
	High staff turnover, low staff base	Enhance relationships with management and workforce, better training and commitment to investment in the workforce, other retention initiatives such as performance incentives
	Increased recruitment cost	See above
	Building overoccupancy	Regular occupancy forecasting
Financial WLCC risks	Increases in interest rates	Simulation techniques to forecast likely implications of a deviation in the rate – apportion funding based on this information
	Lack of future investment in capital	
	Increased financial liabilities	Effective financial management protocols, minimise the need for loans, etc.
	Disposal risk	Seek out potential asset purchasers well in advance of decommissioning
	External finance risk	See above
	Unexpected building, plant and equipment obsolescence (technological)	Extend warranties on critical equipment

12.9 Summary

Developing strategies for identifying and responding to risk throughout the WLCC has been the key theme throughout this book. It is, however, during the operational stage of the building that this is all the more important. The WLCC of any building is weighted significantly towards operational costs, so dealing effectively with these risks will minimise the risk to WLCC. These operational risk strategies are all the more important in PFI/PPP projects where key risks lie in contractor delivery and operational costs. In the next chapter we will move on to look at modelling WLCC during the operational stage. By using a case study example, we will look at both the application of WLCC in an existing building, and also consider how a design and construction phase WLCC model can be updated and revised during the operational stage.

References

Holmes, A. (1994) In *Chartered Institute of Building (CIOB) Handbook of Facility Management* (ed. A.L. Speddin). Longman Group Limited, Harlow.

IFMA (undated) *What is Facilities Management*? International Facilities Management Association, Houston.

IPD (2002) *The Occupiers Property Databank International Total Occupancy Cost Code.* Investment Property Databank, London.

NHS Estates (1996) *Re-engineering the Facilities Management Service.* Health Facilities Note 16. The Stationery Office, London.

The Stationery Office (2000) *Examining the Value for Money of Deals Under the Private Finance Initiative.* The Stationery Office, London.

13 Whole Life-cycle Costing during Operational Stage

13.1 Introduction

The application of the WLCC techniques we have discussed throughout this book can provide the building occupier with a great deal of information about the economic efficiency of the building. The operational stage of the building is the period where WLCCs are most focused and using models to monitor these costs over a period of time provides useful information to the stakeholders. Depending upon the individual building under analysis, operational stage WLCC models fall broadly into two categories:

- *New WLCC models*: WLCC models for existing buildings where no prior model was developed during the design stage
- *Existing WLCC models*: WLCC models that were developed during the design stage and are subsequently revised and updated during the operational stage.

As we have mentioned in previous chapters of this book, the application of WLCC techniques during the operational stage has not been as widespread as for models used during the design stage. One of the drivers behind the desire to develop methodologies for operational stage WLCC modelling is the PFI/PPP contract. The emphasis in these types of contract is ensuring adequate risk transfer to the private sectors whilst still ensuring value for money and quality service delivery throughout the contracted operational phase of the building. Beyond simple benchmarking and key performance measurement, the ability to measure the quality and value of services within the building is difficult. This is where WLCC techniques are most powerful. They enable the analyst to examine the economic performance of the building in context, by providing information on all the operational costs. The building occupiers can then examine the relationship between costs and make future decisions based upon this advice.

This chapter will discuss, by means of case study examples, how to develop and implement a WLCC model for both of the scenarios described. We shall also investigate some of the data problems that may be experienced when handling WLCC models, and how these may be overcome. Again, the emphasis within this chapter is on risk. The examples in this chapter will demonstrate some of the risk modelling techniques that were discussed in Chapter 5.

13.2 Operational stage WLCC models

Whole life-cycle costing techniques are, as we have discussed in previous sections of this book, used primarily as a decision making tool during the preconstruction phase of the life of a building. In other words, they are used to optimise the design decisions made and inform the likely cost implications of these decisions over the whole life. It is apparent though that when the operational phase of the building begins, the information obtained from the WLCC analysis is rarely used to inform future decision making. In fact, it could be observed that in many cases the results of the WLCC analysis are simply discarded and not referred to subsequently.

Whole life costing exercises must always include forecasts of long-term, maintenance, asset replacement, operational and financial costs, as these form the basis of the design decision making process. But during the operational phase of the building, the building owner will also want to know how the building is performing economically in comparison with the original WLCC forecasts. This is important because operational costs account for the largest proportion of whole life costs.

13.3 The importance of operational costs in WLCC

At the core of WLCC philosophy are operational costs. As many design decisions are based solely on initial capital costs and not future operational costs, it is quite easy to understand why WLCC has not developed to the extent envisaged. To many stakeholders, the idea of incorporating the operational costs into the overall decision making process will become more widely accepted when guarantees can be made about the accuracy of the forecasts. This is really the basis behind the integration of risk in WLCC, providing information on the uncertainty element of WLCC. Forecasting operational costs has an inherent risk, and since the ratio of operation costs to capital costs is very high, the risk is also high. It has been argued that the very basis of WLCC is essentially concerned with optimising the operating costs of physical assets to minimum (Woodward 1997). Accordingly, accurate forecasting of operating costs is essential to minimise the total WLCC of the building. Recent research has suggested that for every £1 spent on capital costs, £50 is spent on maintenance costs and £200 is spent on operational costs (Royal Academy of Engineering 1999). It is therefore clear that the accuracy of WLCC is strongly correlated to the accuracy of the operational cost forecasts.

13.4 Conceptualising the WLCC model

As we discussed in the examples given in Chapter 12, there are many costs that could be considered. These costs will vary depending upon the type of building and function; the generic WLCC centres will normally still hold though, and these can be used to establish the basic framework of the model. One thing to bear in mind is that the best WLCC models are tailored specifically to the building under analysis. By using knowledge from the design and construction teams and information from the building owner, a WLCC that suits the particular needs and requirements of the organisation can be produced. Generic models may neglect useful cost data and possibly invalidate the accuracy of the results.

First and foremost, however, before beginning to develop the WLCC model, we need to define if the model required is:

- A new WLCC model for operational stage analysis
- An existing WLCC model to be updated.

This simple distinction has a major bearing upon how the model will be constructed or revised. Naturally, if the WLCC model is already in existence, a great deal of the work will have already been achieved. Developing the structure of the model is probably the most difficult part of the overall process, and this can be not only time consuming but also costly. There is therefore a distinct advantage to the analyst if an existing model is already available. On the positive side, however, developing the WLCC design model in the first place is a fairly large task, so developing a WLCC model for an existing building does not require any modelling of the capital cost centre, as this is known and not an uncertain value.

13.5 Existing WLCC models

An existing WLCC model (i.e. a model that was prepared prior to construction of the building) will detail not only the capital cost items as we mentioned earlier, but also forecasts for each of the operational cost centres that were identified in the examples given in Chapter 12. These values are uncertain, and therefore are treated as stochastic values. As the life of the building progresses, actual cost-in-use data will become available. This information is useful in that it not only allows us to update the existing model by replacing stochastic forecasts with certain values, but also allows us to compare the actual performance of the building with the forecasted costs. Simply recomputing the WLCC by running a new simulation to include the new cost data in replacement of the relevant annual stochastic values can produce a simple percentiles graph such as the one shown in Fig. 13.1; we can compare the revised WLCC distribution with the original.

Naturally, the analyst will be able to carry out other statistical tests on the results, such as measuring the sensitivities of each of the cost centres on the WLCC for each year, as shown in Fig. 13.2. Ensuring that tests like this are conducted yearly also helps to monitor costs and identify if any particular cost centres are affecting the overall WLCC. What is important, however, is that an appropriate format for collecting this data is established.

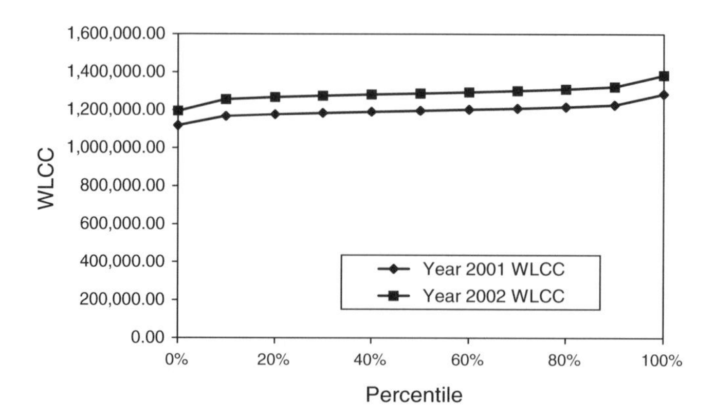

Fig. 13.1 Comparing WLCC forecasts with revised data.

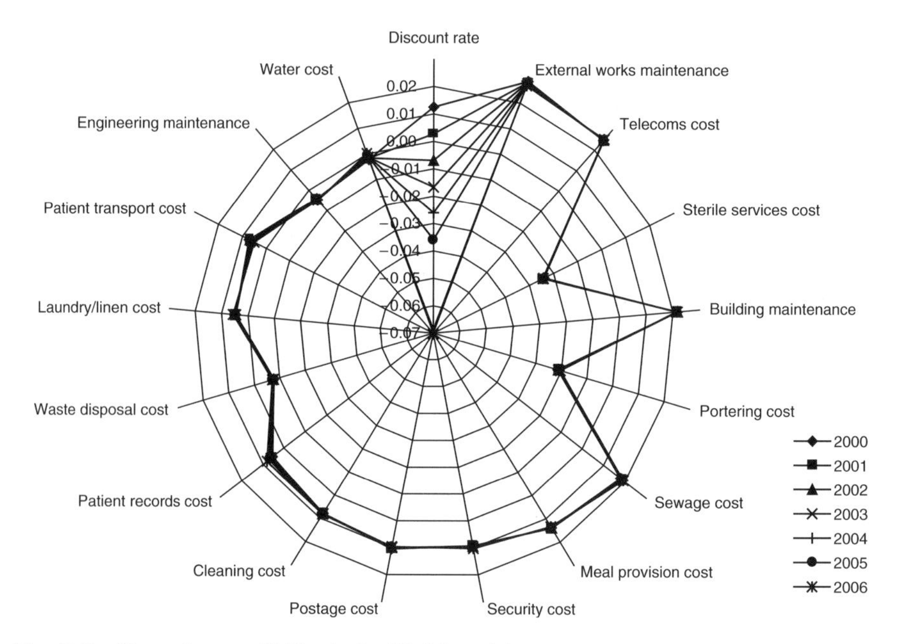

Fig. 13.2 Measuring sensitivities in the WLCC model.

13.6 Recording WLCC data during occupancy stage

It is throughout the occupancy stage of the building that actual cost-in-use data becomes available to the analyst and this can prove to be very useful for updating the model and revising the original. Although we know that this data exists during the occupancy stage, the mechanisms must be in place to ensure that the data is recorded in an accurate and systematic way. It is therefore important that along with the WLCC model, a framework for the collation of this data also exists. The problems of implementing WLCC models are nearly always attributed to concerns over the inaccuracy of the forecasts, due to lack of quality data. With respect to the element of inaccuracy, this is directly related to the absence of adequate cost data collection mechanisms. If, for example, all building occupiers were required to submit annual running cost profiles, the risk associated with LCC techniques could be significantly reduced (Bird 1987). In fact, White (1991) argues the case for 'performance profiles' and in particular highlights again the requirements for a universal construction data information system. The sheer complexity of many existing WLCC models lends little to practical application and in many cases, if not the majority, the lack of available good quality data prohibits further use. In terms of the practitioners, they need to be willing to promote clients and building occupiers into adopting a more holistic approach to running cost control so that data collection mechanisms can be put in placed to aid and assist all those requiring LCC cost profiles.

A model would be developed whereby a software application that works alongside a WLCC model to record the cost-in-use data is being devised. The application, known as the 'logbook', is in effect a repository for recording the design, capital investment, maintenance, operational and decommissioning decision making process throughout the whole life of the building or critical structure. The logbook empowers the user with ability to record these decisions, so that valuable knowledge can be elicited on cost, performance and economic viability of an asset.

Furthermore, the logbook provides the means for basing future decisions on the most up-to-date cost and performance information. Perhaps more importantly, the logbook serves as the vehicle by which to encourage the building/critical structure owner to remain focused on whole life costs during the occupancy stage, and establish the procedures required to accurately record cost-in-use data. This project is timely in that it addresses the major problem remaining in WLCC methodologies – obtaining accurate cost-in-use data. It may be the case that future research will look towards developing a system that can harmonise data recorded via 'logbooks' to one central WLCC cost data source similar to the BCIS/BMI (Building Maintenance Information). The basic process of the logbook is shown in Fig. 13.3 where the

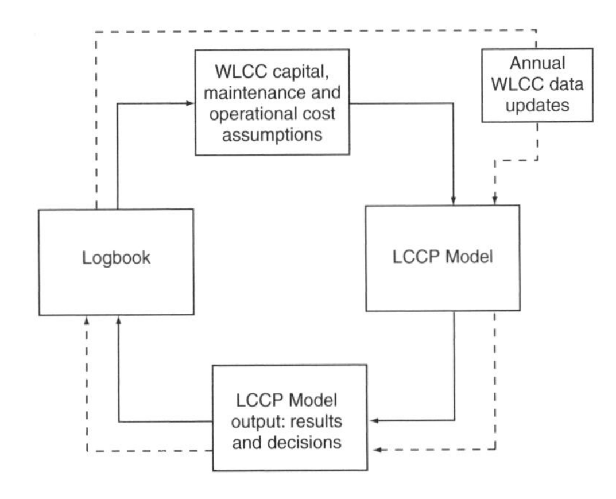

Fig. 13.3 The logbook data collection framework.

process of updating the model with new cost-in-use data and WLCC forecast revision can be appreciated.

However, even where such a formal framework for recording costs is not in place, it is still common for historical records of certain data sets to be available. In many larger buildings historical records of cost and consumption of energy are accessible. This data is very useful in developing a WLCC model for an existing building. In the next section, a case study example is presented on how to utilise this data to develop risk-integrated energy WLCC forecasts.

13.7 Forecasting energy WLCC: a case study

In this section, an example is given of how some of the theoretical techniques we have reviewed in earlier chapters of the book could be implemented into a WLCC model. Historical consumption of electricity cost data from an NHS hospital was extracted from records. The modelling procedure to convert this information into WLCC forecasts essentially formed two distinct stages. The first stage involved the selection of the forecasting technique that was most appropriate. The forecasting methods investigated include naïve methods, exponential smoothing methods and an extension to exponential smoothing methods, the Holt-Winters method. The second stage discusses the risk associated with the forecasting process and proposes a technique for translating the forecasts (known as deterministic assumptions) into stochastic values. These stochastic values can then be used as inputs to a simulation model.

13.7.1 Stage one: Time series forecasting of electricity cost

For example purposes only, the forecast period is set for four years. Due to data availability and time lags in receiving historical cost data, the basis of the forecasting procedure would be historical data over the period April 1997 to December 1999. Therefore the period of WLCC forecasting would be up until the year 2003. The hospital building used in this study is the tower block of an NHS trust teaching hospital. Three electricity supply substations service the main ward block building, substation 3, substation 4 and substations 3 and 4 high voltage supplies, which source the supply from a combination of in-house produced electricity and external supply from a local power company.

The first task was to elicit the monthly meter readings from each substation, in kW. To convert the readings into unit cost a moving average was used as the unit cost of electricity each month varied. Figure 13.4 shows the moving average of unit electricity cost over a 12-month period.

The cost data was then exported into a forecasting software package, CB predictor, which was used to generate forecasts for the period January 2000 to December 2003. CB predictor uses four components – level, trend, seasonality and error – to analyse the historical data and then produce the specified forecast (Decisioneering 2000). The software generates up to eight different forecasts using four non-seasonal and four seasonal forecasting

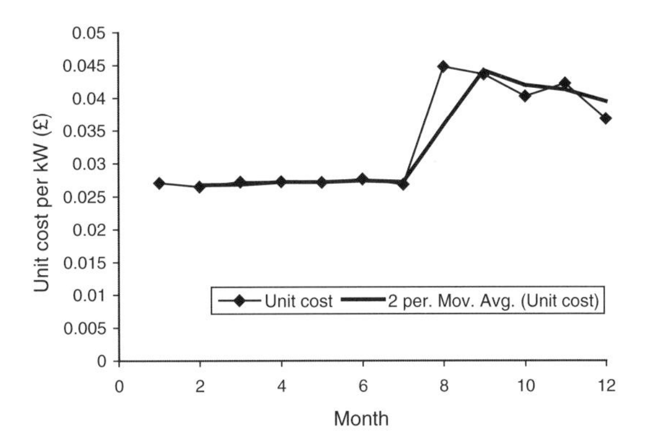

Fig. 13.4 Moving average of unit cost of electricity (kW) for 1999.

methods. The non-seasonal methods available are single moving average, double moving average, single exponential smoothing and Holt's double exponential smoothing. The four seasonal methods available are seasonal additive smoothing, seasonal multiplicative smoothing, Holt-Winters additive seasonal smoothing and Holt-Winters multiplicative smoothing. Data pre-analysis revealed that it would be likely that the seasonal methods would most accurately forecast the future costs, but a comparison of all eight methods for each substation is presented in Tables 13.1, 13.2 and 13.3, which also give the goodness-of-fit parameters of each forecasting method. An independent energy management specialist at the NHS trust examined the results, and after inspection it was found that the forecasts for substations 3 and 4 using the Holt-Winters multiplicative method were satisfactory. However, for the substation 3 and 4 high voltage, the expert cast doubt over the validity of the forecasts produced by the Holt-Winters additive method so a lower ranked method (Holt-Winters multiplicative) was selected in preference. Figures 13.5, 13.6 and 13.7 show the results of the selected forecasting procedure, showing the fit and forecast for each substation and the observed linear trend.

13.7.2 Stage two: Stochastic modelling of annual forecasts

The next stage of the process is to use the results generated in the first stage and transform the forecasts into probabilistic values, which can then subsequently be used as stochastic inputs into a WLCC model. However, in modelling the forecasts as probabilistic values, knowledge is required about the conditions surrounding the variables and the most likely probability distribution that the forecasts belong to. This can prove difficult but the selection of an appropriate probability distribution is very important as it can have a significant effect upon any simulation work that is carried out in the future to calculate the whole life-cycle cost. Therefore, to establish the most likely distribution that the forecasts would come from, a database of historical energy costs for over 450 NHS acute care trusts in England and Wales was analysed, for the periods 1996–97, 1997–98 and 1998–99.

Table 13.1 Forecast methods for substation 3.

Methods	Rank	RMSE	MAD	MAPE	Durbin-Watson	Theil's-U	Periods	Alpha	Beta	Gamma
Double exponential smoothing	7	3200.6	2747.2	18.958	2.528	0.681		0.168	0.999	
Double moving average	8	3317.8	2927.2	20.569	2.818	0.523	9			
Holt-Winters additive	3	2298.6	1665.1	11.595	3.252	0.481		0.041	0.999	0.001
Holt-Winters multiplicative	**1**	**2285.8**	**1559.6**	**10.69**	**3.204**	**0.489**		**0.001**	**0.001**	**0.001**
Seasonal additive	4	2408	1772.6	11.86	3.092	0.536		0.107		0.001
Seasonal multiplicative	2	2298.3	1633.1	10.633	3.106	0.543		0.012		0.001
Single exponential smoothing	5	2676.4	2312.8	15.178	2.834	0.593		0.059		
Single moving average	6	2798.3	2390.4	15.992	2.932	0.558	9			

RMSE = root mean square error; MAD = mean absolute deviation; MAPE = mean absolute percentage error.

Table 13.2 Forecast methods for substation 4.

Methods	Rank	RMSE	MAD	MAPE	Durbin-Watson	Theil's-U	Periods	Alpha	Beta	Gamma
Double exponential smoothing	7	2604.7	2142	17.178	2.695	0.636		0.144	0.999	
Double moving average	8	2885.7	2512.2	20.402	2.735	0.544	9			
Holt-Winters additive	3	1994.4	1586.7	12.611	3.209	0.515		0.046	0.999	0.001
Holt-Winters multiplicative	**1**	**1892.8**	**1302.6**	**10.283**	**3.385**	**0.492**		**0.001**	**0.001**	**0.001**
Seasonal additive	4	2059	1589.6	12.337	3.14	0.543		0.115		0.001
Seasonal multiplicative	2	1924.2	1387.5	10.512	3.217	0.534		0.001		0.001
Single exponential smoothing	5	2386.4	1997.9	15.175	2.774	0.613		0.099		
Single moving average	6	2435.1	2074.2	16.211	2.888	0.56	9			

RMSE = root mean square error; MAD = mean absolute deviation; MAPE = mean absolute percentage error.

Table 13.3 Forecast methods for substations 3 and 4 high voltage.

Methods	Rank	RMSE	MAD	MAPE	Durbin-Watson	Theil's-U	Periods	Alpha	Beta	Gamma
Double exponential smoothing	8	8980	7427.7	30.833	2.079	0.903		0.486	0.398	
Double moving average	7	8228.7	7037.8	27.365	1.796	0.812	7			
Holt-Winters additive	2	6096.2	4468.6	18.633	2.256	0.637		0.04	0.999	0.001
Holt-Winters multiplicative	**3**	**6273.4**	**4017.2**	**16.591**	**2.379**	**0.613**		**0.001**	**0.257**	**0.001**
Seasonal additive	4	6333.7	4925	19.972	2.372	0.692		0.197		0.001
Seasonal multiplicative	1	5981.7	4348.4	17.082	2.526	0.662		0.067		0.001
Single exponential smoothing	6	7266.2	5217	19.476	2.013	0.854		0.208		
Single moving average	5	6781.8	5120.4	20.61	1.882	0.788	7			

RMSE = root mean square error; MAD = mean absolute deviation; MAPE = mean absolute percentage error.

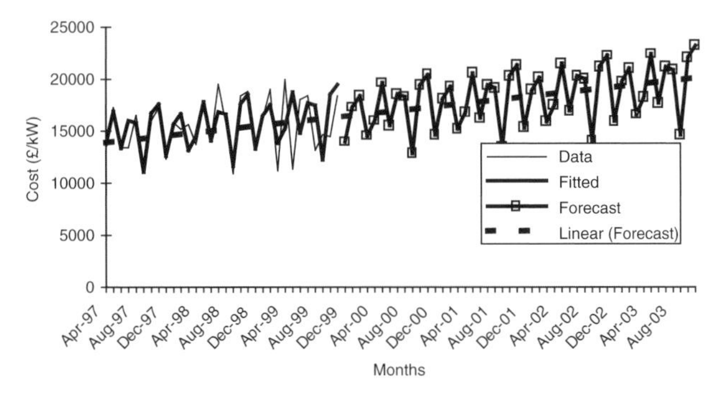

Fig. 13.5 Forecast for substation 3.

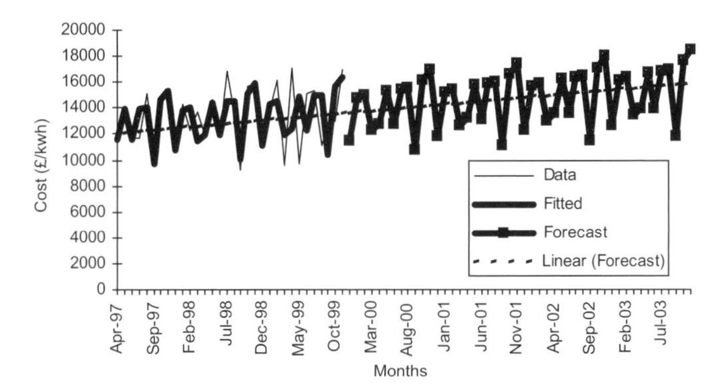

Fig. 13.6 Forecast for substation 4.

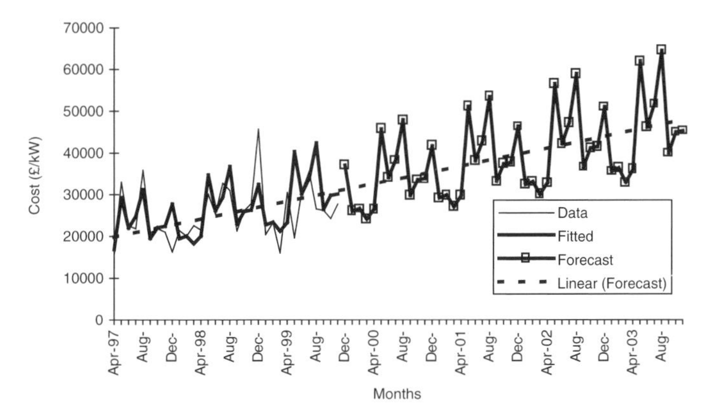

Fig. 13.7 Forecast for substations 3 and 4 high voltage supplies.

The first stage was to remove samples from the sets that contained energy cost data on non-hospital sites. This was performed because the data was collated at trust level, i.e. for all buildings within a trust's estate, not for individual buildings. Non-hospital sites were removed because they do not reflect the true energy costs of acute care hospitals and would thus distort distribution fitting later on in the research. Once this had been performed, eliciting a set of observed data that had, on aggregate, an approximately equal mean floor area to that of the ward block building in the university teaching hospital, then reduced the sample further. Similarly, building characteristics such as gross heated volume were used to further reduce the sample. After consultation with practitioners, the final data sample was reduced to 52 sets to eliminate data sets that had missing or erroneous data. Predata analysis revealed that the variation in electricity costs was not very high and therefore the final data set could be deemed as being homogeneous. All final data sets exhibited similar characteristics to that of the main ward block building used as the basis for the study.

A probability distribution fitting programme was then used to transform the data from each year into probability density functions. The purpose of this was to establish any similarities between each year's cost data and also to establish any particular distribution or family of distributions that best represented the cost data. It was found that in this case, for hospital buildings with the same physical characteristics to that of the hospital building used in this study, the Weibull distribution was ranked first in terms of goodness-of-fit for the periods 1996–97 and 1998–99 and ranked second in the period 1997–98, based upon the Anderson-Darling (A-D) goodness-of-fit test.

Equipped with the knowledge that the annual costs of electricity in this hospital were likely to come from the Weibull distribution, the monthly point forecasts generated in strand one of this chapter were then used as normally distributed assumptions in a Monte Carlo simulation model. The purpose of this was to generate a distribution of the annual forecasted cost for the years 2000 to 2003. The A-D test was used to assess the validity of the Weibull distribution in preference to the first ranked distribution and it was found that for all four years, the Weibull distribution was accepted by the A-D test and the error in the Weibull model mean relative to the data sample mean was either 0.01% or 0.02%, indicating a negligible error.

The parameters of these distributions can then be elicited from the software and input directly into the WLCC model. Figure 13.8 shows the results of the simulation for the forecast year 2003. The resulting probability distribution is shown along with probability–probability (P–P) plot and a distribution–function–differences (D–F–D) plot, which basically show in visual format the goodness-of-fit of the distribution.

13.8 New WLCC models for operational stage WLCC analysis

We have discussed in the last section how it is possible to update an existing WLCC and compare the results of actual and forecasted values. However, we now need to look at developing WLCC models for existing buildings where no previous model has been employed.

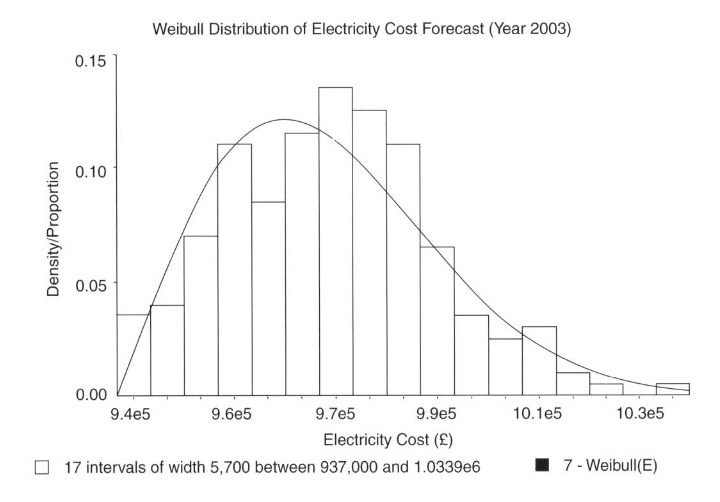

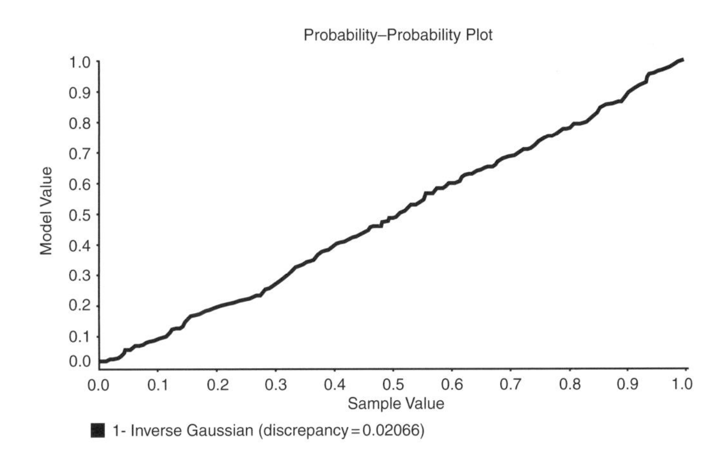

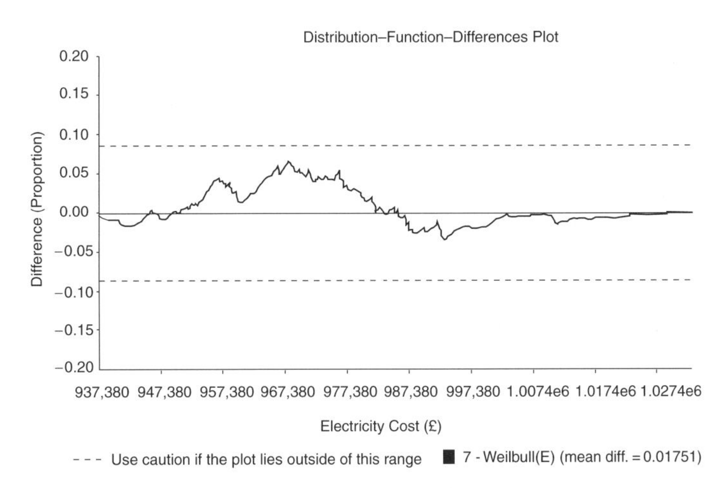

Fig. 13.8 Forecast distribution, probability–probability plot, distribution–function–differences plot and parameters for year 2003.

Depending upon the circumstances of the particular project, developing the WLCC could be fairly straightforward. If the building owners have maintained accurate and regular records of cost-in-use data, then this historical information could be used to generate future WLCC. But what if this information is not available? Is it possible to develop a WLCC model anyway?

The answer to this question is yes, but a thoroughly robust and statistically accurate approach to modelling the operational WLCC must be applied. If little or no data exists, the first stage must be to establish a process for doing this. Once this is in place, then the ability to revise and update the model in the fashion previously described is available. To set up the WLCC model initially, a variety of data sources may be available to the analyst. Before we consider these sources, it is important to be aware of some of the problems that are associated with using data from external sources.

13.8.1 Data handling

In carrying out the WLCC analysis, we need to consider many operational cost items. These will typically include such items as cleaning costs, facilities management provision, security, staff costs in relation to the running of the building, catering provision, etc.; however, much of this data in current formats is not representative of a building, but over an entire estate of buildings (Kirkham *et al.* 2002). This is an idiosyncrasy of many organisations with large capital estates such as schools, prisons, defence organisations and hospital healthcare providers. The point here is that a strategy must be established so we can deal effectively with this data within the WLCC model, and create sample populations within the database underpinned by the most appropriate statistical methodologies.

The treatment of the data used to model operational WLCC also has implications for dealing with the intended stochastic nature of the results of the WLCC model. Probabilistic modelling of data in detailed component WLCC analysis is very different from the techniques that are used in concept stage analyses. Recent work on detailed level WLCC has utilised specialised probabilistic techniques such as Markov chains (generally for deterioration models and service life forecasting) and risk-integrated time series forecasting models (energy cost forecasting and financial planning). Generally, large-scale models use specific sampling techniques to estimate the parameters of probability distributions, which can be used to represent a population more accurately. Naturally, a model that incorporates such a level of detail is time consuming and difficult to implement within a short time scale. It is therefore important to be focused on how the data will be modelled when collecting cost-in-use data

13.8.2 Sources of external data

Having considered the treatment of operational cost data within the model, the analyst has a variety of possible sources of data, as follows.

Quantity surveyors' cost models

Many of the building engineering and surveying journals now produce cost models for new build projects, particularly large-scale projects such as hospitals and government buildings. These cost models will normally include estimates of operational costs, derived from experience of past project. These can be used as the basis for estimating the operational costs in the WLCC analysis. To derive probabilistic estimates of these costs, it is possible to obtain three exclusive cost model estimates and construct a probability distribution from these values.

Three-point estimation

Three-point estimating is one of the most popular methods of representing the forecasted costs. In forecasting terms, a three-point estimate is an approximation of the range of possible costs from a minimum to a maximum, with the most likely cost appropriately located between these two extremes.

It is also a methodology for describing the quantification of risk and the limits of variability of uncertainty that surround forecasts in a format suitable for further, useful, analysis. It should be noted, however, that a clear rationale for obtaining these values is set down and that the data is as accurate as possible; simply specifying three values based on nothing more than guesswork will yield inaccurate and possible misleading results.

Sampling from existing FM cost databases

Many large organisations have now established and maintain facilities management cost databases. These databases record various costs of operating the building on an annualised basis. These databases can be useful for operational cost forecasting at the design stage. Moreover, any organisation that wishes to maintain a WLCC analysis during the occupancy phase should establish such a system to record annual cost-in-use data. This will provide the necessary information to update the model in future months.

Meter readings of similar buildings (for energy costs)

Forecasting energy costs can be achieved easily where historical records of energy consumption are retained in similar buildings. Time series forecasting models can be used where existing data exists; these can prove to be reliable and accurate in forecasting these costs, as we have seen earlier in this chapter. In the absence of this type of data, three-point estimating or sampling can be used as a suitable alternative.

13.9 WLCC for an existing building: a case study of facilities management cost modelling

Whole life-cycle costing has been identified by the NHS as having the valuable ability to provide insight into the economic efficiency of their hospitals. Many new PFI hospitals are designed using WLCC techniques but this is not the case for the majority of the older building stock. NHS Estates collate annual data on facilities management costs from NHS trusts so as to give the government a snapshot of the variance of FM costs through the entire NHS estate. This is generally used to create benchmarks whereby NHS trusts can gauge the economic efficiency of their FM services against national averages.

The nature of this data though poses problems for use in WLCC exercises. As the data is collated at trust level, it represents total spending on FM costs for all buildings within the trust's estate, not for individual buildings that make up the entire trust property portfolio. Most hospital trusts encompass several buildings within the estate and hence the data in its raw form cannot be used to assess WLCC. To facilitate the use of this FM data, it must be transformed to reflect the costs of an individual building within a trust's estate portfolio, not for the estate as whole.

An original data sample of over 450 acute care NHS hospital FM costs was obtained with the cooperation of NHS Estates. However, the data provides FM costs for the entire NHS trust estate, therefore in its original format the data reveals little about the cost of FM services in particular buildings within the estate, and thus cannot be used in a WLCC approach. Therefore, a novel approach to transform the data set into usable cost information was proposed. This is illustrated in Fig. 13.9.

Initially, it was decided that the simplest parameter to use for defining the probability distributions was the gross floor area parameter. In other words, each FM cost is defined as a function of the trust's gross floor area. Initial statistical investigation revealed that the data set would need to be reduced, as primarily the variance and standard deviation of the floor area were too high. Also, the data set included cost data on what are referred to as 'non-hospital sites', these being sites that do not serve the primary functions of an acute care hospital, such as administration buildings and primary care centres. Data sets that included costs on non-hospital sites were removed because they do not reflect the true FM costs of acute care hospitals and would significantly distort the results. The removal of data sets that included these costs reduced the sample to 128 sets. Once this had been performed, the mean of the gross floor area (x) of the data set was calculated. This was not sufficiently approximate to the floor area of the ward block building such that ($-10 \leq x \leq +10$) so two data sets were removed, one at a time, (the minimum and maximum samples) until the final overall data set had on aggregate an approximately equal mean floor area to that of the ward block building used as the case study. This resulted in 52 data sets for the modelling phase.

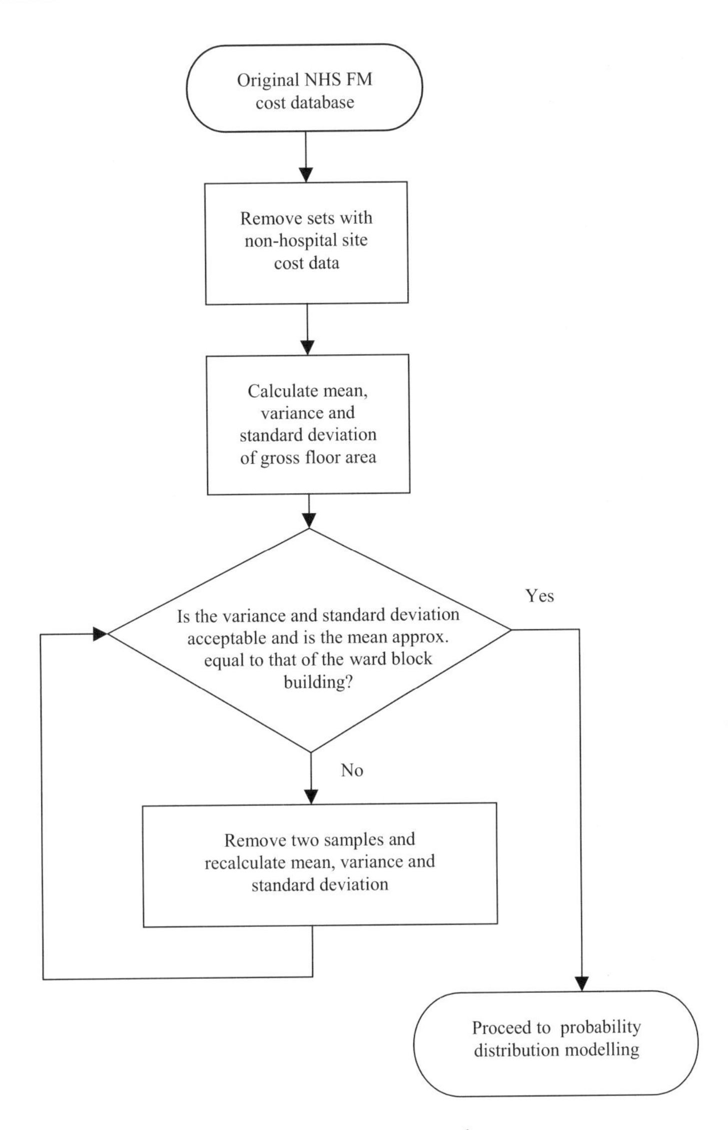

Fig. 13.9 Flow process chart of the data reduction exercise.

Having obtained the sample data, the sets were then statistically analysed for distribution fitting using two software applications, ExpertFit™ and BestFit™, resulting in a probability distribution for each FM cost centre. Testing the distribution against 28 continuous probability distributions and the chi-square goodness-of-fit test validated each distribution. The two packages were used to compare results and identify any ambiguity in first ranked fits. It was found that both packages yielded almost identical results, with the percentage error (δ) falling in the region $0.03 \leq \delta \leq 0.09$ for all distribution fits. It was therefore arbitrarily decided to use the results from the ExpertFit™ software package. The parameters for each of the distributions were elicited, which could then be used in the WLCC model.

13.10 Summary

The applicability of WLCC techniques to existing buildings has rarely been investigated in the past, but the power and applicability of such approaches have been advocated strongly in the examples we have looked at in this chapter. Building up a WLCC model for both of the cases we have described above can be a complex and time-consuming task depending upon a variety of factors such as availability of data, type of project and skill of the analyst. Highly complex WLCCs can be developed as can simple ones; the power of the model is really only limited by the ability of those building and using the model.

It is hoped that this chapter and the following one will provide the reader with some practical examples of how the theory of WLCC and quantitative risk analysis can be combined to create accurate and statistically robust WLCC models.

References

Bird, B. (1987) Costs-in-use: principles in the context of building procurement. *Construction Management and Economics*, **5** Special Issue.

Decisioneering (2000) *CB Predictor user manual*. Decisioneering Inc, Denver, Colorado.

Kirkham, R.J., Boussabaine, A.H. & Awwad, A.H. (2002) Probability distributions of facilities management costs in NHS acute care hospital buildings. *Construction Management and Economics Journal*, **20**, 251–61.

Royal Academy of Engineering (1999) *The Long Term Cost of Owning & Using Buildings*. Royal Academy of Engineering, London.

White, K.H. (1991) Building performance and cost-in-use. *The Structural Engineer*, **69**(7), 148–51.

Woodward, D. (1997), Life-cycle costing – theory, information acquisition and application. *International Journal of Project Management*, **15**(6), 335–44.

14 Whole Life Costing of Building Assets Occupancy: a Case Study

14.1 Introduction

Total occupancy costs are quantified, analysed and presented as part of the investment decision making process in the ever-changing business environment. The need to assess alternatives and make significant investment decisions relating to occupancy costs with limited information may require decision makers to address investment decisions with quantitative processes (Jones 1995).

Total occupancy costs projection is a quantification and presentation of the monetary resources required for running an occupied facility efficiently and effectively. The running costs of buildings of the same age, size, location and general characteristics can differ quite considerably (Dunse & Jones 1998). These differences may be attributed to many factors such as level of maintenance, type of occupancy, attitude of the occupants, etc. The attributes that drive the cost of total occupancy are listed in the occupiers property databank (OPD) international total occupancy cost code; readers are advised to consult the OPD publications for further details.

This in turn has a great influence on the uncertainty of total occupancy costs. In fact, it can be argued that many occupancy cost decisions are made on the basis of uncertain cost information, which may result in a significant number of occupants unable to optimise the use of cost resources over their period of occupation. Therefore, appropriate analytical methods should be developed and used to establish total occupancy costs over an organisation's period of occupation.

An analytical approach is developed in this case study for determining total occupancy cost probability distribution functions (PDFs). These distributions are derived from the expected value and variance of PVs of four cost centres under different cash flow scenarios over the occupancy period.

14.2 Sources of occupancy costs

The OPD international total occupancy cost code has categorised occupancy costs into five broad classes (IPD 2001):

(1) *Real estate occupation costs*
- Rent
- Unitary charge
- Acquisition, disposal and removal
- Taxes
- Parking charges
- Associated facilities cost.

(2) *Adaptation and equipment costs*
- Fit-out and improvements
- Furniture and equipment.

(3) *Building operation costs*
- Insurance
- Internal/external repair and maintenance
- M&E repair and maintenance
- Churn
- Moving out
- Security
- Cleaning
- Waste disposal
- Water and sewerage
- Energy.

(4) *Business support costs*
- Telephones
- Catering
- Reception
- Mail services
- Reprographics
- Transport
- Archiving.

(5) *Occupancy management costs*
- Management
- Facilities
- Project cost.

The above classification is for the purpose of accounting principles rather than for predicting and managing total occupancy costs. From a purely financial perspective total occupancy costs can be defined under four different cost centres: set-up costs; non-controllable occupancy costs; controllable occupancy costs; and disruption costs. The model developed in this case study is based on the assumption that the four occupancy cost centres and the occupancy period are randomly distributed according to one of the theoretical distribution forms. This requires that each cost centre element is treated stochastically and the cash flow (PV) in each year of occupancy is described as a PDF or uncertainty cash flow profile. These PV cash profiles are considered to be variable from one year to another depending on inflation and interest rates. Hence if the PDF or cash flow profile of each occupancy cost centre is known or can be simulated, the total PV of occupancy cost can be estimated using the following equations:

$$PV = S_c + \sum_{i=1}^{n}(R_c + C_c + D_c)/(1 + r)^i \tag{1}$$

Where S_c is the initial set up costs, C_c is non-controllable occupancy costs, R_c running costs and D_c disruption cost relating to occupancy problems. The second term of equation (1) can also be expressed in the following formula:

$$C_{vr} = \sum_{i=1}^{n}(R_c + C_c + D_c) \times P(ri)/(1 + r)^i \tag{2}$$

Where C_{vr} is variable occupancy cost, $P(ri)$ is yearly probability relating to interest rate, r interest rate and i is year of occupancy. Equation (2) expresses the yearly stochastic nature of the occupancy cash flow cost. This equation is useful for carrying out a sensitivity analysis to find out the parameters that influence the occupancy costs. As described above, R_c, C_c, D_c and r are random quantities, and hence, C_{vr} is a function of several random parameters. The composition of R_c, C_c and D_c may have a significant influence on the form of the PDF of yearly and total PV computation. The following sections describe the occupancy cost parameters and functions assumed in this study.

14.3 Set-up costs

The total set-up costs (S_c) for preparing a facility for occupancy over the total period of occupancy are assumed to be fixed and composed of three cost elements:

$$S_c = \sum(A_c + T_c + F_c) \tag{3}$$

Where A_c = agents and legal fees; T_c = stamp duty; and F_c = fitting out cost. The determination of agents' and legal fees is usually very flexible and the methodology used to arrive at this cost can vary from a simple flat fee to a system based on the actual rented area or percentage of a rent payable on an agreed time scale. The guidelines issued by the stamp office and HM Treasury can be used to deduct the stamp duty. Fitting out and removal of internal fittings usually contribute a large proportion of the set-up costs. Usually most of these costs are hidden to tenants. This is mainly due to the fact that through the occupation period reorganisation and staff movements occur due to many internal and external factors, as will be discussed under the disruption costs subheadings. Also tenants may be obliged under the rental contract to pay for clearing and improving the rented space to a reasonable standard. The cost of fitting out and removal of internal fittings is largely dependent on the layout and construction technology used to separate office spaces. Most common layouts include open plan, cellular accommodation and a combination of both. It is envisaged that with the development of innovative materials and construction methods, like mobile wall partitions, the cost of set-up and clearing will be reduced dramatically. Also, fitting out costs depend on the quality and standard of fittings.

14.4 Non-controllable occupancy costs

The non-controllable occupancy costs (C_c) include cost items such as rent, rates and service charge. These costs are assumed to depend on the location, size of the facility, proximity to public transport, age, parking area, type of layout, security, and reception area (Dunse & Jones 1998). Also these costs are not determined by the occupancy of a facility but they are mainly influenced by external factors. Accordingly, C_c can be expressed as:

$$C_c = \sum(C_{re} + C_{ra} + C_{se}) \tag{4}$$

Where:
C_{re} = rent costs
C_{ra} = rates cost
C_{se} = service charges.

These costs tend to vary subjectively according to what the occupants perceive to be the tangible benefits from the rented space in terms of suitability to business, location, access and other potential material benefits. Tenants have little direct influence during the period of occupancy over these costs.

14.5 Controllable occupancy costs

Total running costs (R_c) in a given occupancy period are assumed to be composed of three cost elements:

$$R_c = \sum(C_{ut} + C_{cl} + C_{ma}) \tag{5}$$

Where:
C_{ut} = utilities cost
C_{cl} = cleaning costs
C_{ma} = maintenance/decoration costs (this is only for internal costs).

External maintenance/repair, buildings insurance and security are assumed to be paid for by the landlord and recovered from the occupant through service charges. R_c costs depend to a large extent on the level of usage, management and running of the occupied facility, occupant attitudes to energy-saving measures and other aspects which are deemed important to the efficient running of occupied spaces. These costs can be influenced and optimised through an effective facilities management programme.

14.6 Disruption costs

Disruption costs during a given occupancy period may occur due to several internal and external factors. Among these are absenteeism due to sick building syndrome, and organisational changes (i.e. staff movement from one location to another within an occupied space due to promotion or movement due to a new business environment). This will result in disruption to business activities and productivity loss. The majority of these costs are not considered by analysts, which may have resulted in failure to optimise the real total occupancy cost over the rental period. Here, the cost of disruption to the occupier is assumed to be composed of two elements:

$$C_d = \sum(C_{mv} + C_{ab}) \qquad (6)$$

Where:
C_{mv} = cost related to moving and relocating individuals within the occupied space
C_{ab} = cost related to loss of productivity due to absenteeism because of sick building syndrome and other factors related to the occupied space.

C_{mv} costs are estimated as a function of the rate of movement of individuals in an organisation within the occupied space. This rate is particularly high during the early years of occupancy when occupants are getting accustomed to their new working environment. Also external environmental factors, like rapid changes in the market and economic structure, may accelerate the rate of movement. It was reported that a typical rate of movement could range between 43% and 80% depending on the circumstances of a particular organisation (Facilities 1991). Also the cost of organisational changes due to individual movement has been estimated at between £1000 and £3000 per person per move depending on the size of the organisation and the sophistication of fittings and other requirements for the move (Price Waterhouse 1992). Hence, the cost due to personnel movement can be estimated in several ways. A lump sum can be allocated to cover these costs. This can be distributed either uniformly or unevenly over the rental period. The following formula could also be used to estimate the disruption cost due to personnel movement:

$$C_{mv} = C_{or} \times N_p \times N_v \qquad (7)$$

Where:
C_{or} = estimated cost for organisational change
N_p = number of employees likely to move or change working space over a specific occupation period
N_v = forecasted number of moves for each employee over a specified period of occupancy.

The drawback of equation (7) is that it is not related to any of the factors that influence organisational changes. It may be argued that the more congested a space, the more likely that organisational change will occur. Hence, it might be better that C_{mv} cost should be expressed as a function of the size of floor area of the space occupied by each employee:

$$C_{mv} = \frac{C_{or} \times N_p \times N_v}{A_f} \tag{8}$$

Where A_f = space occupied per employee.

This formula suggests that the greater the space allocated to each employee the cheaper the cost for organisational change. However, equations (7) and (8) do not take into account organisational change due to external factors like the state of the market and economic conditions. Hence, it might be better to model C_{mv} probabilistically to reflect the uncertainty and variability associated with organisational change:

$$C_{mv} = \frac{C_{or} \times N_p \times N_v}{A_f} P(or) \tag{9}$$

Where $P(or)$ = probability that organisational change will occur.

It is expected that in the early stages of occupancy and in volatile market conditions this probability should be very high. Because of the lack of reliable data sources for estimating the parameters in equations (6–9), organisational change could be easily expressed as a percentage rate of the number of employees that could be affected by organisational change. Hence C_{mv} could be estimated from the following equations:

$$N_v = O_r \times N_p \tag{10}$$

$$C_{mv} = N_v \times C_{or} \tag{11}$$

Where:
N_v = number of person moves
O_r = organisational change rate
N_p = number of employees
C_{or} = estimated cost per person per move.

Cost of absenteeism due to rented space problems can be modelled by two cost elements:

$$C_{ab} = \sum C_{sk} + C_{bs} \tag{12}$$

Where:
C_{sk} = cost of absenteeism due to the effect of sick building syndrome on the occupant productivity and performance.

It is estimated that employees could lose up to 1% of their productivity due to sick building syndrome problems (Building 1988; RICS 1991). Therefore, C_{sk} might be computed using the following equation.

$$C_{sk} = C_{ap} \times N_p \times R_{sk} \tag{13}$$

Where:
C_{ap} = average total cost per employee
N_p = number of employees
R_{sk} = percentage of time loss due to sick building syndrome
C_{sk} = estimated cost of sick building syndrome per annum.

The cost associated with general absenteeism (C_{bs}) as a result of space occupation can be estimated using the following expression:

$$C_{bs} = C_{ap} \times N_p \times R_{bs} \tag{14}$$

Where R_{bs} = the percentage rate of absenteeism per annum per employee.

Therefore, by substituting equations (13) and (14) into (12) the annual occupancy cost due to absenteeism can be computed using the following formula:

$$C_{ab} = C_{ap} \times N_p(R_{sk} + R_{bs}) \tag{15}$$

Equations (1–15) incorporate the time value of money into the analysis of occupancy costs using PV concept. These equations are used in simulation software to model occupancy costs and account for the potential variability of the parameters used in determining whole life-cycles of office occupancy. The results of this technique provide probabilistic distribution on the potential whole life-cycle costs of office occupancy.

14.7 Data and methodology

The aim here is to explore the potential of a stochastic approach for the analysis of office cost occupancy, and the application of probability concepts for occupancy cost quantification and interpretation. This type of analysis provides a cost analyst with a method to express and quantify risk involved in occupancy cost variables instead of relying on a single-point forecasting method.

The data used in this research was acquired from office occupancy costs in London. The characteristics of offices included in this study are summarised in Table 14.1. The table shows that four offices were newly built and four were internally refurbished. All offices except one have parking facilities. Eight offices are air-conditioned. Only one office block has conference facilities and another has a restaurant and shop with health club, pool and apartments.

The occupancy data was divided into four cost centres, as shown in Table 14.2. The division was based on the principle that the costs that can be controlled by the occupants should be separated from non-controllable costs, so that optimisation can be easily carried out. Also, disruption costs should be dealt with separately due to their high uncertainty and the complexity of forecasting them.

Table 14.1 The characteristics of studied offices.

Location	Area to let (m^2)	Costs (£/m^2)				Characteristics				Condition		
		Rent	Rates	Services	Fit-out reinstatement	A/C	Raised floor	A/C ducting	Parking	Modern	Façade retention	Refurb.
Millbank	7896				Included	Yes	Yes	No	Yes	No	Yes	No
Millbank	3584	269	129	53	Not included	Yes	No	Yes	Yes	No	No	Yes
Westminster	4366	269	118	64	Not included	No	No	Yes	Yes	No	No	Yes
Trafalgar Square	3808	349	107	70	Not included	Yes	Yes	Yes	Yes	No	Yes	No
Victoria	2043	317	161	60	Not included	Yes	Yes	Yes	Yes	Yes	No	No
Buckingham	6874	376	172	70	Not included	Yes	Yes	Yes	Yes	Yes	No	No
Great George St	529	322	139	43	Not included	Yes	Yes	Yes	No	No	No	Yes
Stage Place	7153	269	126	43	Not included	No	No	No	Yes	No	No	Yes
Vauxhall Bridge	3623	317	161	48	Not included	Yes	Yes	Yes	Yes	Yes	No	No
Victoria	2043	403	139	59	Not included	Yes	Yes	Yes	Yes	Yes	No	No

Table 14.2 Input assumptions.

Assumptions		Distribution	Parameters			Characteristics		
			Min.	Most likely	Max.	Mean	σ/μ	σ
Operational costs	Non-controllable costs							
	1. rent	Triangular	285	326	350	320.3	0.1	3202
	2. rates	Triangular	146	163	180	163	0.1	17
	3. service charge	Triangular	58	65	72	65	0.1	7
	Controllable costs							
	1. utilities	Triangular	29	33	36	32.7	0.1	3.51
	2. cleaning	Triangular	9	10	11	10	0.1	1
	3. decoration	Triangular	0.54	0.6	0.66	0.6	0.1	0.06
Disruption costs	Absenteeism costs							
	1. average total cost per employee	Triangular	24336	27040	29744	27040	0.1	2704
	2. SBS factor	Triangular	0.009	0.01	0.011	0.01	0.1	0.001
	3. absenteeism rate	Triangular	0.035	0.04	0.045	0.04	0.1	0.005
	Churn costs							
	1. churn rate	Triangular	0.55	0.60	0.7	0.62	0.1	0.08
	2. churn cost	Triangular	2800	3000	3500	3100	0.1	360.5
	Total disruption cost							
	1. Churn	Weibull	Loc = 1528; scale = 374.7; shape = 2.74			1908	0.07	126.15
	2. Absenteeism	Weibull	Loc = 1107; scale = 293; shape = 3.6			1370	0.06	77.0

The model used to represent input data follows the three-point estimating procedure (triangular distribution). This method requires the analyser to set a lower and an upper limit to a most likely base estimate for every occupancy cost item, according to the most optimistic and pessimistic conditions, respectively. The degree of variability is indicative of the level of certainty about the occupancy costs. The uncertainty of occupancy cost variables mean that they can be treated as stochastic variables. Monte Carlo simulation can be applied to determine the distribution of PV, given that the probability distribution of each variable is known. The probability distributions of the occupancy cost parameter can be determined from experience of similar occupancy costs in approximately similar conditions. The triangular distribution was selected for modelling the occupancy cost input variables. The rationale behind the selection of triangular distribution is based on the following assumptions (Fente *et al.* 2000; Law & Kelton 1991; Wall 1997).

(1) Occupancy costs must have a specified lower and upper limit. Beyond these limits, the occupancy costs cannot assume any values. Therefore, this assumption infers that selected distribution should be close-ended.
(2) The occupancy cost parameters may have any value within the defined upper and lower limits. This assumption infers that the distributions for the cost input variables should be continuous.
(3) The triangular distribution function parameters can be easily estimated by experts (Chau 1995).
(4) Occupancy costs tend to vary greatly depending on several parameters, as discussed above. This assumption suggests that skewness must be expected in the distributions that represent the occupancy cost input parameters.
(5) Cost probability distributions should have a convex shape rather than concave (Back *et al.* 2000).
(6) It has been suggested by many authors that triangular distribution is appropriate for modelling cost-related data probabilistically (Back *et al.* 2000).

The parameters of the triangular probability-density function can be estimated using expert subjective judgement (Chau 1995) or from historical data using moment matching, maximum likelihood estimation, and least-squares fit of the cumulative distribution function (AbouRizk *et al.* 1994). In any case, whatever the form of the distribution that represents the input variables, an estimate of these variables can be obtained through the mean-variance method. In this study the mean of the distribution function of the input variables is assumed to be the estimated value of the base case. The coefficient of variance of the input variables is assumed to be 0.1, so that the standard deviation will be:

$$\sigma = 0.1\,\mu \qquad (16)$$

If an input variable has a high uncertainty the coefficient of variance of this variable can increase depending on the perceived uncertainty associated with the variable.

The most likely values in Table 14.2 were extracted from actual cases. These values along with the assumption made in equation (16) were used to estimate the parameters of the probability distribution function of the input variables. Figure 14.1 shows the process used to develop the simulation model. The process starts with the definition of the input variables assumptions. Due to the lack of data and stochastic nature of equations (10), (11) and 15, a subsimulation model was used to simulate the absenteeism cost and cost related to organisational changes. Goodness-of-fit tests were used to select the best probability distribution function that represents these two subcost centres. The selected distributions, along with their characteristics and goodness-of-fit tests, are shown in Table 14.2. These two distributions are then used as an input to simulate the PV cost due to disruption over the period of ten years.

Monte Carlo simulation is used to generate the distribution of possible PVs. It takes samples from the input variable distributions and evaluates the corresponding PV that is a function of these variables. The process is repeated for 1000 iterations and the resulting PV1, PV2,..., PVn are used to obtain the cumulative distribution of PV. Obtaining PV occupancy costs in this way is subject to estimation error resulting from sampling error, and inappropriate

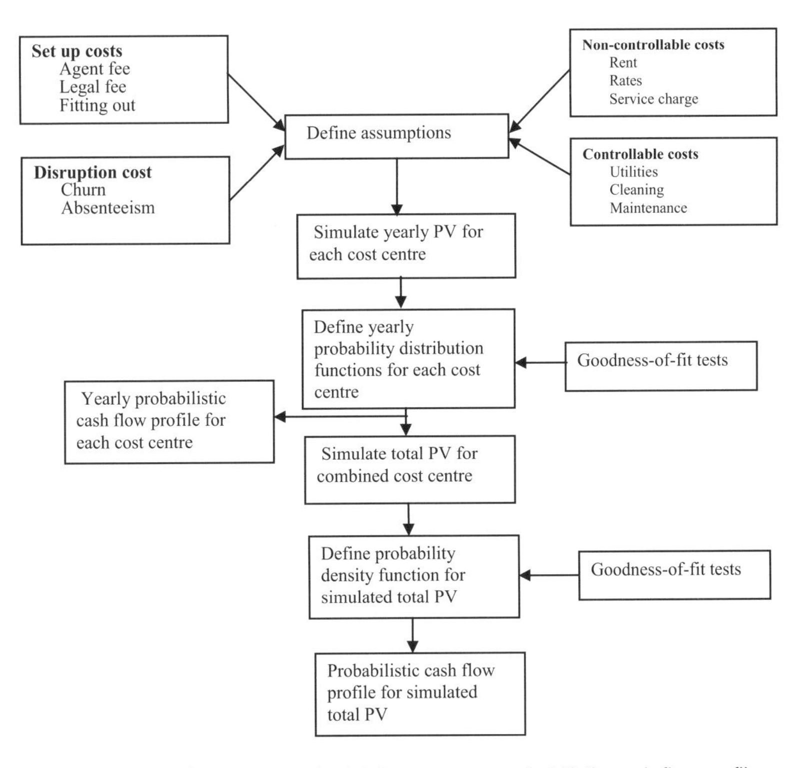

Fig. 14.1 Process used to generate the total occupancy probabilistic cash flow profiles.

discount and inflation rates. Therefore, the reliability of the generated distribution has to be assessed. In this work three goodness-of-fit tests were employed.

The first approach uses the Kolmogorov test, which checks if the observed data could have originated from a theoretical distribution with the estimated parameters. It does not require grouping of data and no interval specification is required. It also does not detect tail discrepancies very well. A value less than 0.03 indicates a good fit.

The second approach utilises the chi-square goodness-of-fit selection method. This test is a formal comparison of the input data histogram with the fitted distribution. It gauges the general accuracy by breaking down the distribution into areas (classes) of equal probability and compares the data points within each area to the number of expected data points. Hence, the choice of the number and size of class intervals greatly influences the results of the chi-square test. To ensure the validity of the test the number of class intervals and the number of samples should be above 5 and 50 respectively. A value greater than 0.5 indicates a close fit.

The last approach uses the Anderson-Darling goodness-of-fit test. This test closely resembles the Kolmogorov test, except that it weights the differences between the two distributions at their tails greater than at their mid-ranges. It is used to find a better fit at the extreme tails of distributions. Hence it is used to detect discrepancies in the tails of studied distributions. A value less than 1.5 indicates a close fit.

14.8 Results and analysis

Tables 14.3–14.5 summarise the results of matching the forecasted occupancy costs to a theoretical distribution function. The matching was based on using the three goodness-of-fit tests described previously. A total of 15 continuous distributions were assessed. Those which best represent the PV of each occupancy cost, along with their characteristics, are shown in Tables 14.3–14.5. Table 14.3 shows that beta and Weibull distribution are ranked first for controllable costs. But lognormal distribution was found to be the best fit for the total PV. What is noticeable from Table 14.3 is that the coefficient of variation and standard deviation of these distributions is very small. This could be attributed to the fact that the occupiers have a good in-house cost control system.

The total PV for controllable costs was found to fit lognormal distribution. This suggests that most of the total PV values will occur near the mean value, as shown in Fig. 14.2. This also suggests that the total PV due to controllable costs cannot fall below the lower limit of zero but may increase to any value without limit. This is the same phenomenon as real estate prices, which show positive skewness since property values cannot become negative.

Table 14.4 shows the theoretical distributions that best fit the forecasted non-controllable occupancy costs. Weibull, normal and beta distribution were

Table 14.3 Characteristics of best-fit distributions for controllable occupancy costs.

Distribution characteristics	Controllable costs										
	Year 1	Year 2	Year 3	Year 4	Year 5	Year 6	Year 7	Year 8	Year 9	Year 10	Total NPV
Type	Beta	Weibull	Weibull	Weibull	Beta	Beta	Weibull	Beta	Beta	Beta	Lognormal
μ	43.3	43.3	43.3	43.3	43.3	43.3	43.3	43.3	43.3	43.3	432.67
σ	1.48	1.47	1.5	1.5	1.5	1.5	1.5	1.5	1.49	1.48	4.83
ν	0.03	0.03	0.03	0.03	0.03	0.03	0.03	0.03	0.03	0.03	0.01
Location		38.48	38.70	38.68			39.06				
Shape		3.61	3.38	3.41			3.07				
Scale	52.96	5.31	5.09	5.11	51.1	52.59	4.71	52.98	54.49	53.6	
α	155.7				130	149.6		154.5	173.1	163.58	
β	34.88				23.62	23.26		34.67	44.9	39.08	

Table 14.4 Characteristics of best-fit distributions for non-controllable occupancy costs.

Distribution characteristics	Non-controllable costs										
	Year 1	Year 2	Year 3	Year 4	Year 5	Year 6	Year 7	Year 8	Year 9	Year 10	Total NPV
Type	Weibull	Normal	Beta	Beta	Weibull	Beta	Weibull	Beta	Normal	Weibull	Normal
μ	447	448.8	419.5	392	366.4	342.4	320	299	279.5	261.2	3575.6
σ	28.4	94.28	87.05	81.14	74.04	71.5	65.2	60.9	57.7	53.8	220.16
ν	0.06	0.21	0.21	0.21	0.21	0.20	0.21	0.20	0.20	0.21	0.06
Location	344.6				151.03		102.09			91.45	
Shape	4.04				3.19		3.72			3.5	
Scale	112.7		798.22	765.4	240.46	650	241.38	628.9		188.68	
α			10.49	10.87		10.34		12.18			
β			9.47	10.36		9.29		13.43			

Table 14.5 Characteristics of best-fit distributions for disruption occupancy costs.

Distribution characteristics	Disruption costs										
	Year 1	Year 2	Year 3	Year 4	Year 5	Year 6	Year 7	Year 8	Year 9	Year 10	Total NPV
Type	Lognormal	Gamma	Lognormal	Gamma	Gamma	Lognormal	Gamma	Beta	Beta	Gamma	Gamma
μ	3066	2864	2676	2502	2339	2185	2043	1909	1784	1668	23053
σ	137	131.4	121.7	111.4	108.9	99.9	93.2	87.7	82.2	75.6	337.3
ν	0.04	0.05	0.05	0.04	0.05	0.05	0.05	0.05	0.05	0.05	0.01
Location		1893		1474	1320		1117			915.4	19549.5
Shape		51		78.7	84.29		94.2			92	96.9
Scale		19		13	12.1		9.84	4417.8	4161.6	8.18	36.17
α								254.2	253.8		
β								333.67	337.8		

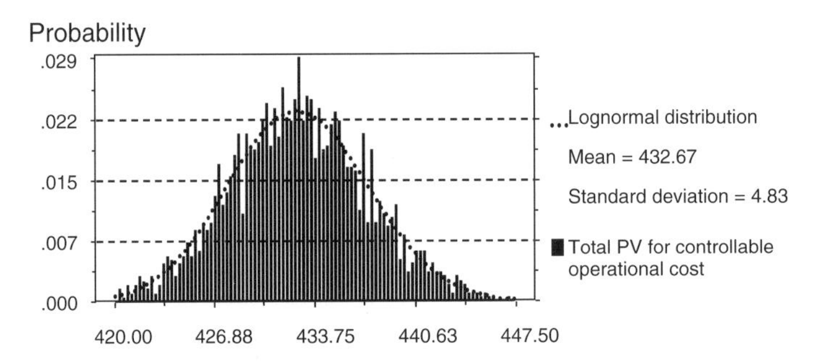

Fig. 14.2 Best fit probability distribution for controllable occupancy costs.

found to be the best fit for this cost centre. The coefficient of variation is very small whereas the standard deviation has a wide range of values. This could be attributed to the uncertainty and volatility nature of this cost centre. Also, occupants have little control over rate charges, taxes, etc. The total PV for non-controllable costs fits normal distribution as shown in Fig. 14.3. This suggests that total PV for this cost centre is most likely to be equal to the mean of the distribution; also it would be likely to be above or below the mean (symmetrical about the mean). But it is more likely to be close to the mean than far away from it. Approximately 68% of PV values will be within one standard deviation of the mean.

The theoretical distributions that best represent disruption occupancy costs are listed in Table 14.5 and the best fit for the total disruption cost is shown in Fig. 14.4. Here it was found that lognormal, gamma and beta distribution best represent the forecasted disruption costs. Again, it is noteworthy that the coefficient of variation is very small, whereas the PV standard deviation of each year tends to fluctuate from year to year. In Table 14.5 six columns of PV were represented by gamma distribution and two were lognormally distributed. It may be argued that because gamma distribution applies to a wide range of physical quantities and is similar to lognormal distribution, it may be used to fit the PV forecast for years 1, 3 and 6. Years 8 and 9 were best fitted to beta distribution. This suggests that the PV value for these two years is a random value between 0 and the scale of the PDF (approximately £4400 per person).

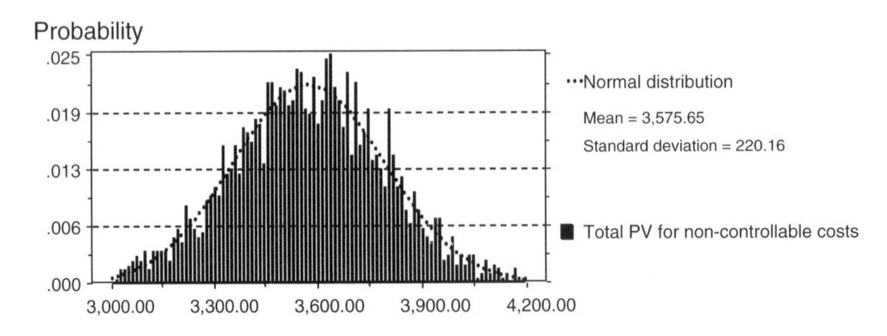

Fig. 14.3 Best fit probability distribution for non-controllable occupancy costs.

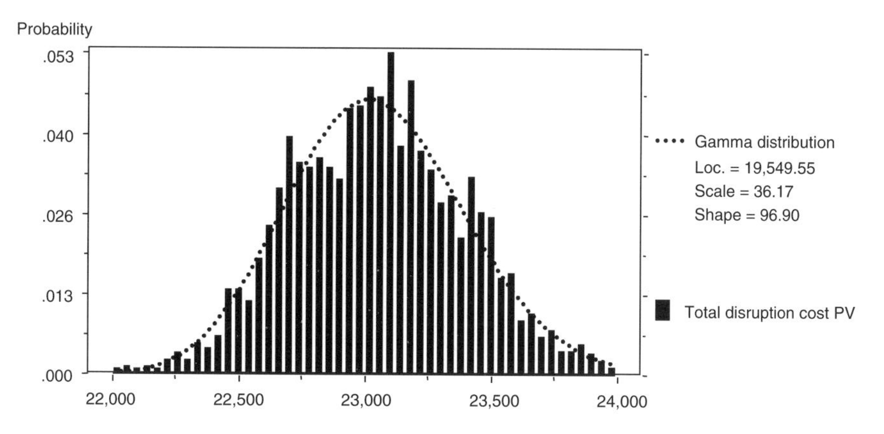

Fig. 14.4 Best fit probability distributions for disruption occupancy costs.

Figure 14.5 shows that the total combined PV (which includes all three cost centres) is slightly positively skewed, rather than normally (symmetrically) distributed. PV costs exhibit this trend because the PV values cannot fall below the lower limit of zero but may increase. These PV values cannot be negative, for example cost for rent charges will never be negative. Also, the logistic distribution for the total combined PV suggests that the occupancy cost grows (increases) as a function of time cost.

Figures 14.6–14.8 show the cash flow profile of each cost centre over the study period. Disruption and controllable cost curves show that the range between PV at 0 and 100 percentile is small whereas the range of non-controllable costs is widely spread. This confirms and validates the characteristics of the generated distributions.

On the basis of goodness-of-fit tests and the data set used here, the log-normal, normal, beta, gamma, logistic and Weibull distributions are found

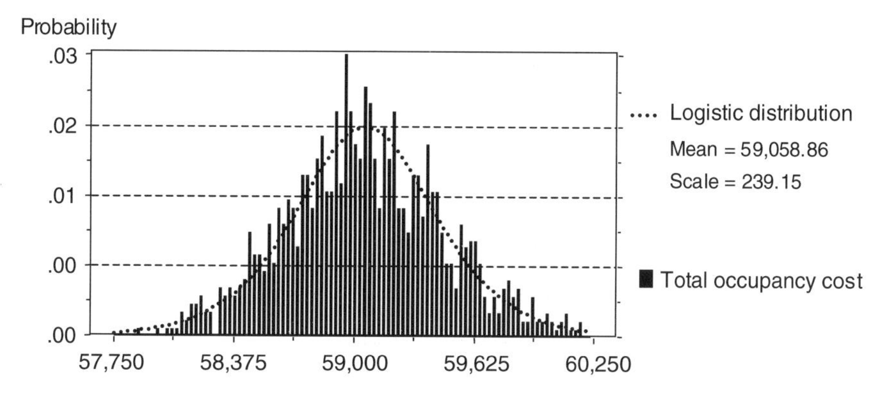

Fig. 14.5 Best fit probability distribution for total occupancy cost.

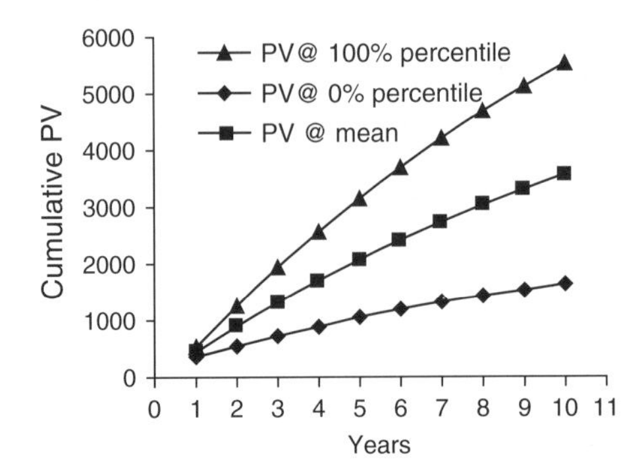

Fig. 14.6 Cash flow profile for non-controllable occupancy cost.

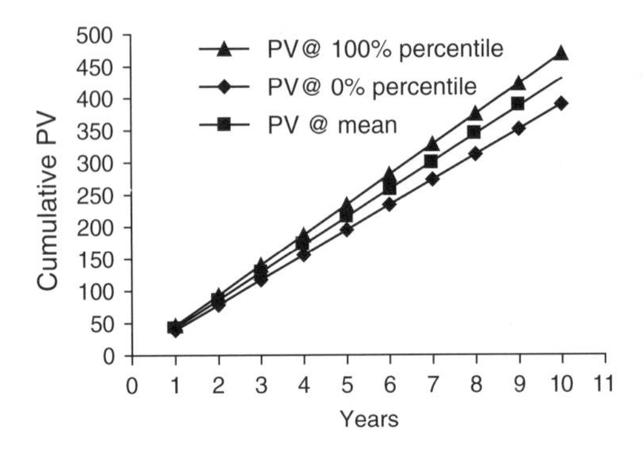

Fig. 14.7 Cash flow profile for controllable occupancy cost.

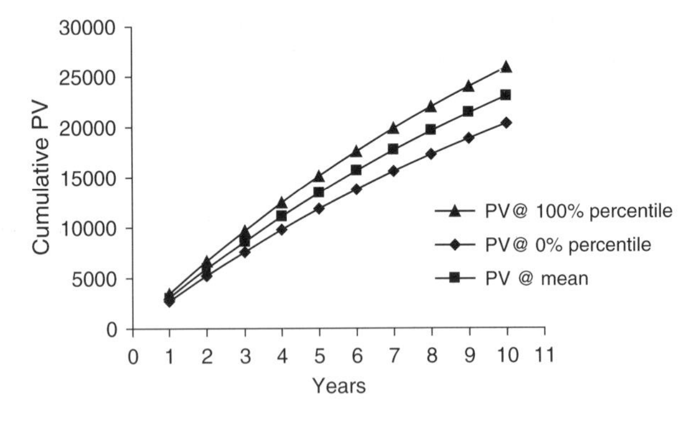

Fig. 14.8 Cash flow profile for disruption occupancy costs.

to represent occupancy cost. At the very least this result suggests that the use of these distributions should be subjected to closer scrutiny and further investigation. This case study has demonstrated a new approach to occupancy cost analysis. When using a range of forecasting methods for office occupancy cost, the occupancy costs are assumed to be random variables rather than known variables. As a result, a probable distribution of total occupancy cost is forecasted as compared to a single-point forecast that would be generated in a deterministic occupancy cost analysis. The resulting probability distributions enable stakeholders to define occupancy cost values that can be associated with prescribed levels of certainty, thus allowing a quantification of the exposure to financial risk. Interested parties can represent the occupancy costs as a probability distribution and make appropriate statistical inferences to guide financial decision making.

14.9 Summary

This case study has introduced an innovative concept for occupancy cost analysis. Using WLCC and risk analysis techniques provides a wide range of values that are dependent upon the level of confidence that is desired for occupancy cost decisions. Based on the type of analysis presented in this chapter, facilities occupants can more accurately assess the potential life-cycle cost associated with office occupancy at a chosen level of risk. By using the probabilistic method, the expected occupancy cost ranges may reflect accurately the actual life cost of occupancy.

Results show that the yearly PV of office occupancy cost can be modelled by lognormal, normal, beta, gamma and Weibull distributions. The total PV of each occupancy cost centre is found to be best fitted to gamma for disruption costs, lognormal for controllable costs and normal for non-controllable costs. The combined total occupancy cost is found to be best fitted to logistic distribution. The generated probability distributions can help facility occupants to make strategic decisions based on analysis results that truly reflect the inherent risks and costs related to rented spaces. The results of this work can be used at an early stage of the renting process of space to carry out what-if scenarios.

References

AbouRizk, S.M., Halpin, D. & Wilson, J.R. (1994) Fitting beta distributions based on sample data. *Journal of Construction Engineering and Management*, **120**(2), 288–305.

Back, E., Boles, W. & Fry, G. (2000) Defining triangular probability distributions from historical cost data. *Journal of Construction Engineering and Management*, **126**(1), 29–37.

Building (1988) Sick office – off Sick. *Building*, 11 March, 66–7.

Chau, K.W. (1995) The validity of triangular distribution assumption in Monte Carlo simulation of construction costs: empirical evidence from Hong Kong. *Construction Management and Economics*, **13**(1), 15–21.

Dunse, N. & Jones, C. (1998) A hedonic price model of office rents. *Journal of Property Valuation and Investment*, **16**(3), 297–312.

Facilities (1991) Churn rates. *Facilities*, **9**(9).

Fente, J., Schexnayder, A. & Knutsom, K. (2000) Defining a probability distribution function for construction simulation. *Journal of Construction Engineering and Management*, **126**(3), 234–41.

IPD (2001) *The Occupiers Property Databank International Total Occupancy Cost Code.* Investment Property Databank, London.

Jones, C. (1995) An economic basis for the analysis and prediction of local office property markets. *Journal of Property Valuation and Investment*, **13**, 16–30.

Law, A. & Kelton, D. (1991) *Simulation Modelling and Analysis.* McGraw-Hill International Editions, New York.

Price Waterhouse (1992) *Office Occupancy Costs.* Price Waterhouse, London.

RICS (1991) Keeping a check on sick buildings. *Weekly Chartered Surveyor*, 10 October, 28–9.

Wall, D. (1997) Distribution and correlations in Monte Carlo simulation. *Construction Management and Economics*, **15**, 241–58.

Index